Geometry

Study Guide and Intervention Workbook

New York, New York
Columbus, Ohio
Chicago, Illinois
Woodland Hills, California

To the Student:

This *Study Guide and Intervention Workbook* gives you additional examples and problems for the concept exercises in each lesson. The exercises are designed to aid your study of geometry by reinforcing important mathematical skills needed to succeed in the everyday world. The material is organized by chapter and lesson, with two study guide worksheets for every lesson in *Glencoe Geometry*.

To the Teacher:

Answers to each worksheet are found in *Glencoe Geometry Chapter Resource Masters* and also in the Teacher Wraparound Edition of *Glencoe Geometry*.

The McGraw·Hill Companies

Copyright © by The McGraw-Hill Companies, Inc. All rights reserved. Printed in the United States of America. Except as permitted under the United States Copyright Act, no part of this book may be reproduced in any form, electronic or mechanical, including photocopy, recording, or any information storage or retrieval system, without permission in writing from the publisher.

Send all inquiries to:
Glencoe/McGraw-Hill
8787 Orion Place
Columbus, OH 43240-4027

ISBN: 978-0-07-860191-0
MHID: 0-07-860191-6

Geometry
Study Guide and Intervention Workbook

16 17 045 11 10 09

Contents

NAME ______________________ DATE __________ PERIOD _____

1-1 Study Guide and Intervention

Points, Lines, and Planes

Name Points, Lines, and Planes In geometry, a **point** is a location, a **line** contains points, and a **plane** is a flat surface that contains points and lines. If points are on the same line, they are **collinear**. If points on are the same plane, they are **coplanar**.

Example **Use the figure to name each of the following.**

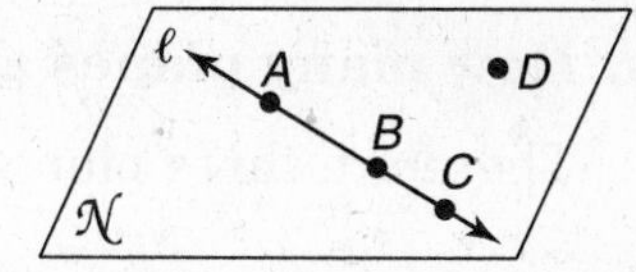

a. a line containing point *A*

The line can be named as ℓ. Also, any two of the three points on the line can be used to name it.

$\overleftrightarrow{AB}$, $\overleftrightarrow{AC}$, or $\overleftrightarrow{BC}$

b. a plane containing point *D*

The plane can be named as plane $\mathcal{N}$ or can be named using three noncollinear points in the plane, such as plane *ABD,* plane *ACD,* and so on.

Exercises

Refer to the figure.

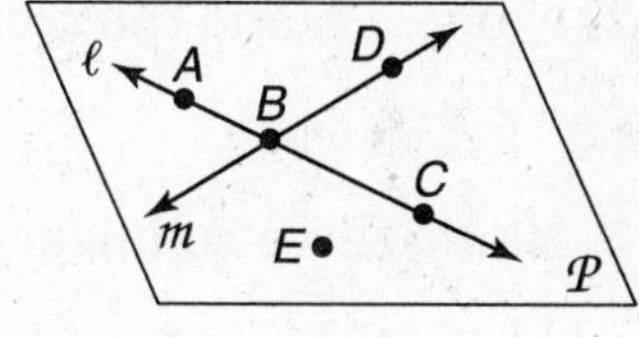

1. Name a line that contains point *A*.

L

2. What is another name for line *m*?

$\overleftrightarrow{BD}$

3. Name a point not on $\overleftrightarrow{AC}$.

E or D

4. Name the intersection of $\overleftrightarrow{AC}$ and $\overleftrightarrow{DB}$.

B

5. Name a point not on line ℓ or line *m*.

E

Draw and label a plane *Q* for each relationship.

6. $\overleftrightarrow{AB}$ is in plane *Q*.

7. $\overleftrightarrow{ST}$ intersects $\overleftrightarrow{AB}$ at *P*.

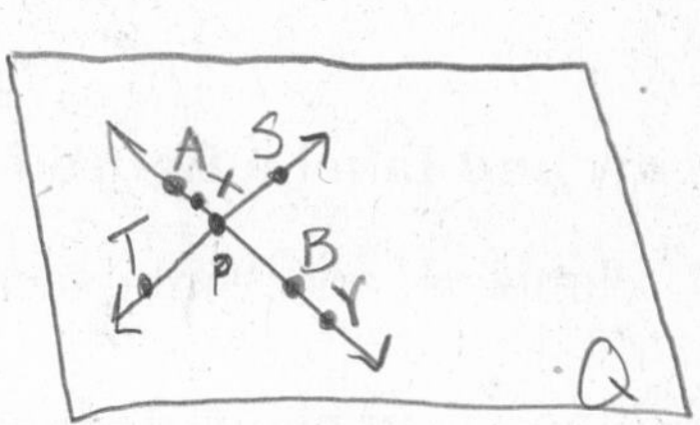

8. Point *X* is collinear with points *A* and *P*.

9. Point *Y* is not collinear with points *T* and *P*.

10. Line ℓ contains points *X* and *Y*.

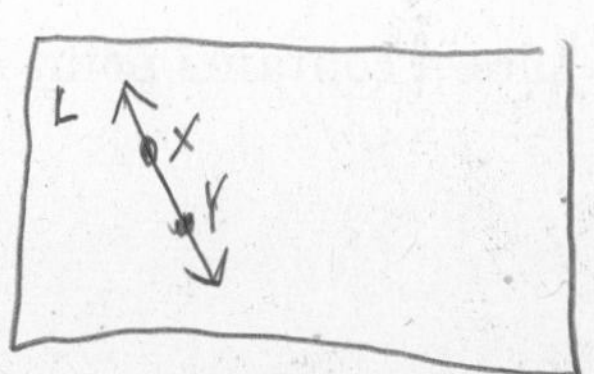

Lesson 1-1

NAME ______________________________ DATE ____________ PERIOD _____

1-1 Study Guide and Intervention *(continued)*

Points, Lines, and Planes

Points, Lines, and Planes in Space **Space** is a boundless, three-dimensional set of all points. It contains lines and planes.

Example

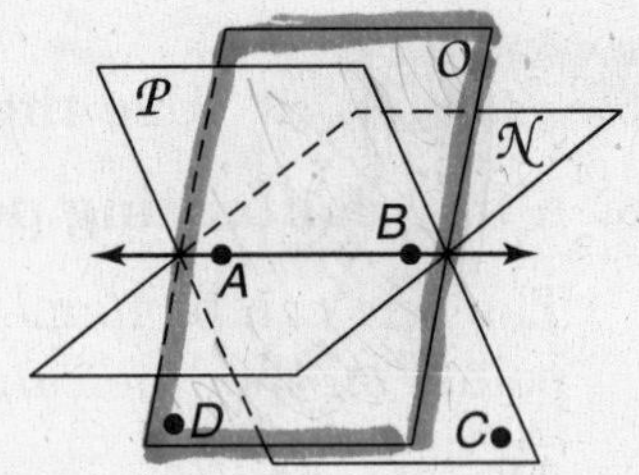

a. How many planes appear in the figure?

There are three planes: plane $\mathcal{N}$, plane $\mathcal{O}$, and plane $\mathcal{P}$.

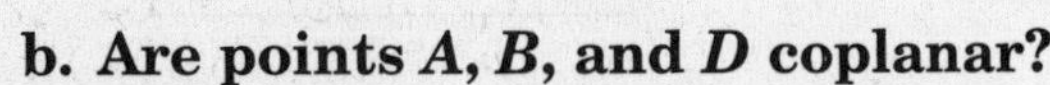

b. Are points *A*, *B*, and *D* coplanar?

Yes. They are contained in plane $\mathcal{O}$.

Exercises

Refer to the figure.

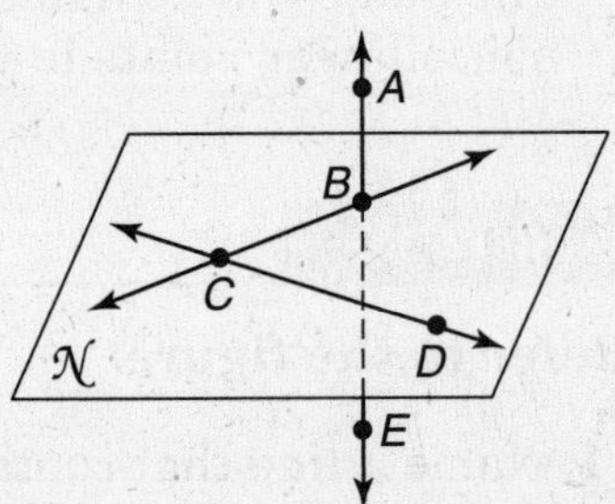

1. Name a line that is not contained in plane $\mathcal{N}$.

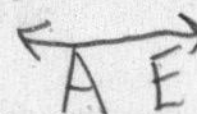

2. Name a plane that contains point *B*.

3. Name three collinear points.

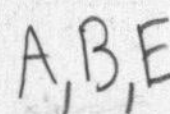

Refer to the figure.

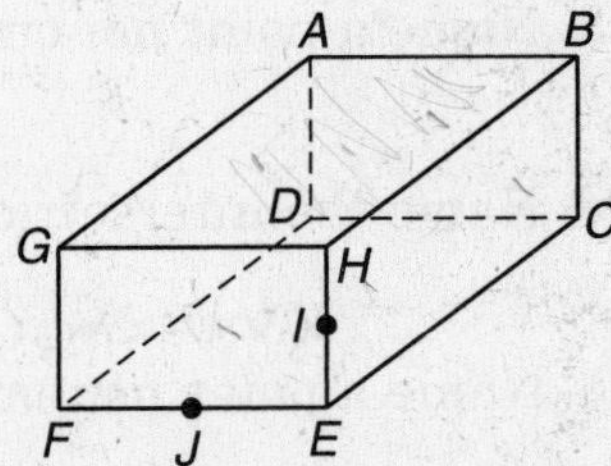

4. How many planes are shown in the figure?

5. Are points *B*, *E*, *G*, and *H* coplanar? Explain.

No, because they lie on different planes

6. Name a point coplanar with *D*, *C*, and *E*.

Draw and label a figure for each relationship.

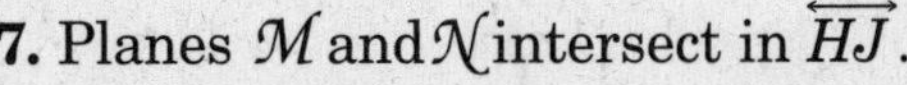

7. Planes $\mathcal{M}$ and $\mathcal{N}$ intersect in $\overleftrightarrow{HJ}$.

8. Line *r* is in plane $\mathcal{N}$, line *s* is in plane $\mathcal{M}$, and lines *r* and *s* intersect at point *J*.

9. Line *t* contains point *H* and line *t* does not lie in plane $\mathcal{M}$ or plane $\mathcal{N}$.

1-2 Study Guide and Intervention

Linear Measure and Precision

Measure Line Segments A part of a line between two endpoints is called a **line segment**. The lengths of $\overline{MN}$ and $\overline{RS}$ are written as MN and RS. When you measure a segment, the precision of the measurement is half of the smallest unit on the ruler.

Example 1 **Find the length of $\overline{MN}$.**

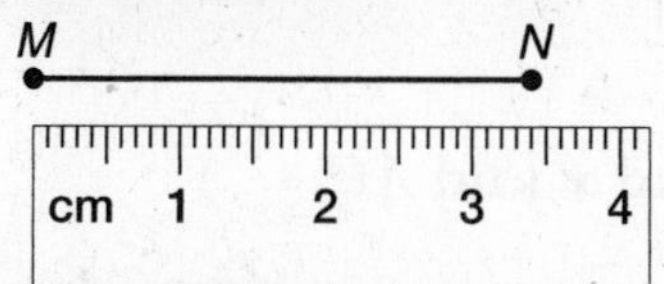

The long marks are centimeters, and the shorter marks are millimeters. The length of $\overline{MN}$ is 3.4 centimeters. The measurement is accurate to within 0.5 millimeter, so $\overline{MN}$ is between 3.35 centimeters and 3.45 centimeters long.

Example 2 **Find the length of $\overline{RS}$.**

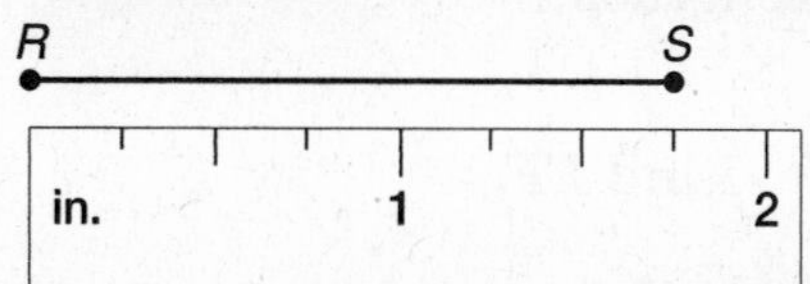

The long marks are inches and the short marks are quarter inches. The length of $\overline{RS}$ is about $1\frac{3}{4}$ inches. The measurement is accurate to within one half of a quarter inch, or $\frac{1}{8}$ inch, so $\overline{RS}$ is between $1\frac{5}{8}$ inches and $1\frac{7}{8}$ inches long.

Exercises

Find the length of each line segment or object.

1. A B

cm 1 2 3

2.

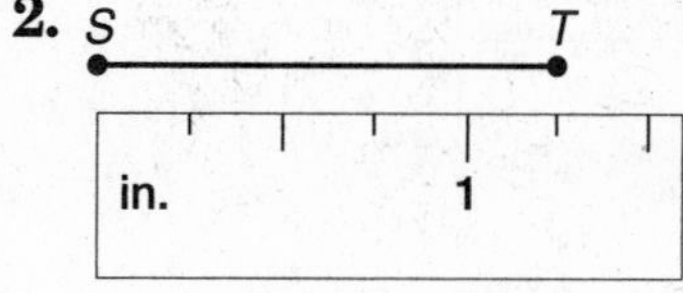

3.

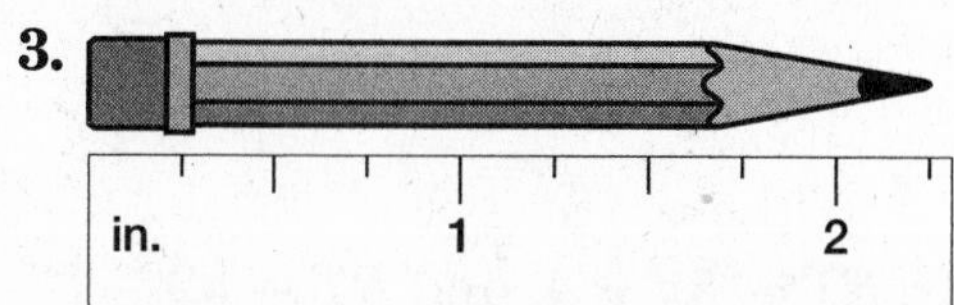

4.

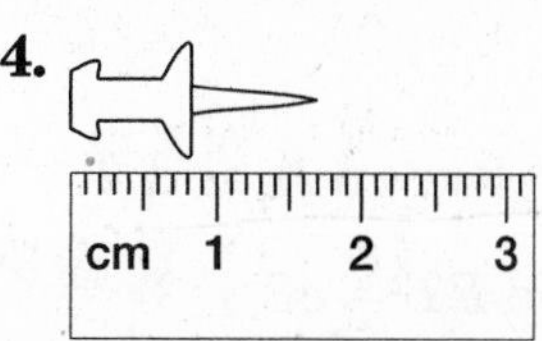

Find the precision for each measurement.

5. 10 in.

6. 32 mm

7. 44 cm

8. 2 ft

9. 3.5 mm

10. $2\frac{1}{2}$ yd

NAME ______________________ DATE __________ PERIOD _____

1-2 Study Guide and Intervention *(continued)*

Linear Measure and Precision

Calculate Measures On $\overleftrightarrow{PQ}$, to say that point M is between points P and Q means P, Q, and M are collinear and $PM + MQ = PQ$.

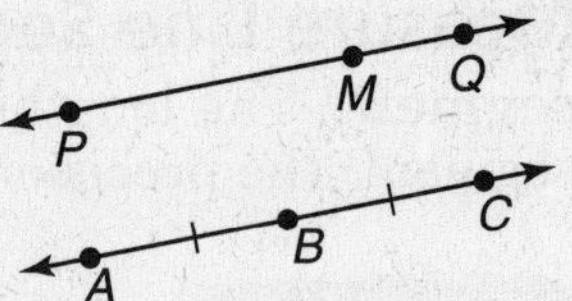

On $\overleftrightarrow{AC}$, $AB = BC = 3$ cm. We can say that the segments are **congruent**, or $\overline{AB} \cong \overline{BC}$. Slashes on the figure indicate which segments are congruent.

Example 1 **Find *EF*.**

1.2 cm 1.9 cm
E D F

Calculate EF by adding ED and DF.

$ED + DF = EF$
$1.2 + 1.9 = EF$
$3.1 = EF$

Therefore, $\overline{EF}$ is 3.1 centimeters long.

Example 2 **Find x and AC.**

$2x + 5$
A x B $2x$ C

B is between A and C.

$AB + BC = AC$
$x + 2x = 2x + 5$
$3x = 2x + 5$
$x = 5$
$AC = 2x + 5 = 2(5) + 5 = 15$

Exercises

Find the measurement of each segment. Assume that the art is not drawn to scale.

1. $\overline{RT}$

2.0 cm 2.5 cm
R S T

2. $\overline{BC}$

6 in.
A $2\frac{3}{4}$ in. B C

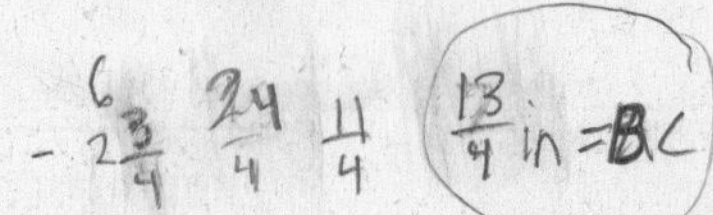

3. $\overline{XZ}$

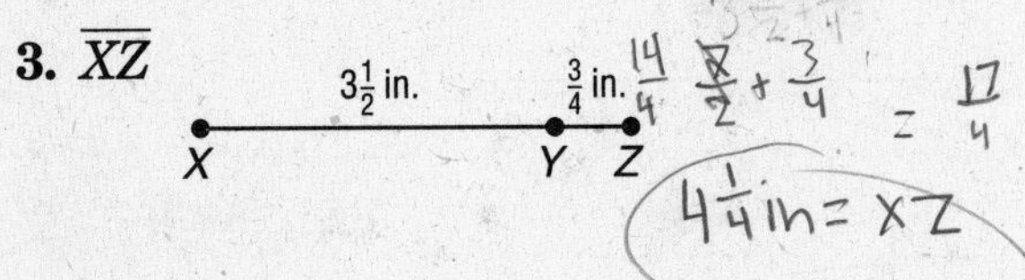

4. $\overline{WX}$

6 cm
W X Y

Find x and RS if S is between R and T.

5. $RS = 5x$, $ST = 3x$, and $RT = 48$.

6. $RS = 2x$, $ST = 5x + 4$, and $RT = 32$.

7. $RS = 6x$, $ST = 12$, and $RT = 72$.

8. $RS = 4x$, $\overline{RS} \cong \overline{ST}$, and $RT = 24$.

Use the figures to determine whether each pair of segments is congruent.

9. $\overline{AB}$ and $\overline{CD}$

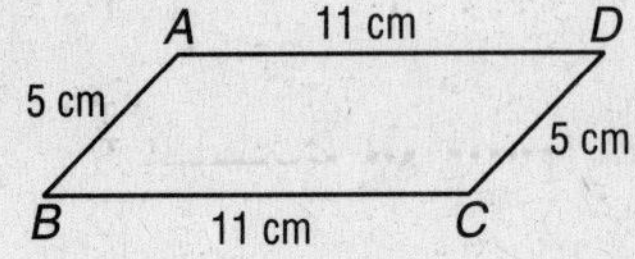

10. $\overline{XY}$ and $\overline{YZ}$

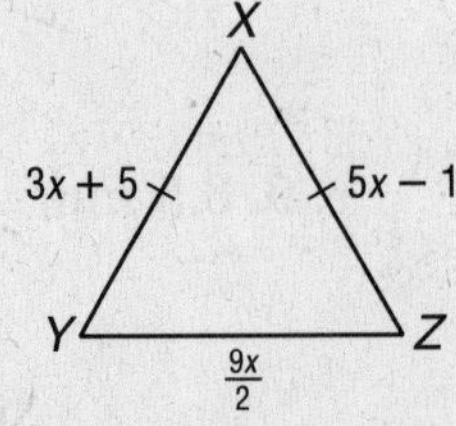

© Glencoe/McGraw-Hill

1-3 Study Guide and Intervention

Distance and Midpoints

Distance Between Two Points

Distance on a Number Line	Distance in the Coordinate Plane
A B (points on a number line at coordinates *a* and *b*) $AB = \lvert b - a \rvert$ or $\lvert a - b \rvert$	Pythagorean Theorem: $a^2 + b^2 = c^2$ Distance Formula: $d = \sqrt{(x_2 - x_1)^2 + (y_2 - y_1)^2}$ 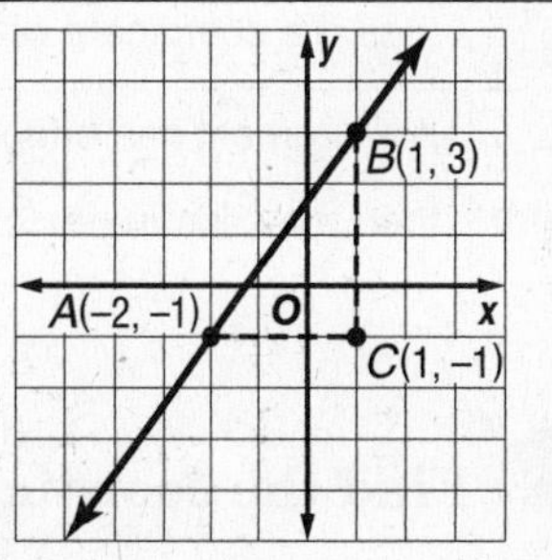

Example 1 **Find *AB*.**

Number line: A at −4, B at 2; −5 −4 −3 −2 −1 0 1 2 3

$AB = \lvert(-4) - 2\rvert$
$= \lvert -6 \rvert$
$= 6$

Example 2 **Find the distance between $A(-2, -1)$ and $B(1, 3)$.**

Pythagorean Theorem

$(AB)^2 = (AC)^2 + (BC)^2$
$(AB)^2 = (3)^2 + (4)^2$
$(AB)^2 = 25$
$AB = \sqrt{25}$
$= 5$

Distance Formula

$d = \sqrt{(x_2 - x_1)^2 + (y_2 - y_1)^2}$
$AB = \sqrt{(1 - (-2))^2 + (3 - (-1))^2}$
$AB = \sqrt{(3)^2 + (4)^2}$
$= \sqrt{25}$
$= 5$

Exercises

Use the number line to find each measure.

Number line: A B C DE F G; −10 −8 −6 −4 −2 0 2 4 6 8

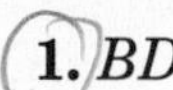

1. *BD*
2. *DG*
3. *AF*
4. *EF*
5. *BG*
6. *AG*
7. *BE*
8. *DE*

Use the Pythagorean Theorem to find the distance between each pair of points.

9. $A(0, 0)$, $B(6, 8)$
10. $R(-2, 3)$, $S(3, 15)$
11. $M(1, -2)$, $N(9, 13)$
12. $E(-12, 2)$, $F(-9, 6)$

Use the Distance Formula to find the distance between each pair of points.

13. $A(0, 0)$, $B(15, 20)$
14. $O(-12, 0)$, $P(-8, 3)$
15. $C(11, -12)$, $D(6, 2)$
16. $E(-2, 10)$, $F(-4, 3)$

1-3 Study Guide and Intervention *(continued)*

Distance and Midpoints

Midpoint of a Segment

Midpoint on a Number Line	If the coordinates of the endpoints of a segment are a and b, then the coordinate of the midpoint of the segment is $\frac{a+b}{2}$.
Midpoint on a Coordinate Plane	If a segment has endpoints with coordinates (x_1, y_1) and (x_2, y_2), then the coordinates of the midpoint of the segment are $\left(\frac{x_1+x_2}{2}, \frac{y_1+y_2}{2}\right)$.

Example 1 **Find the coordinate of the midpoint of $\overline{PQ}$.**

The coordinates of P and Q are -3 and 1.

If M is the midpoint of $\overline{PQ}$, then the coordinate of M is $\frac{-3+1}{2} = \frac{-2}{2}$ or -1.

Example 2 **M is the midpoint of $\overline{PQ}$ for $P(-2, 4)$ and $Q(4, 1)$. Find the coordinates of M.**

$$M = \left(\frac{x_1+x_2}{2}, \frac{y_1+y_2}{2}\right) = \left(\frac{-2+4}{2}, \frac{4+1}{2}\right) \text{ or } (1, 2.5)$$

Exercises

Use the number line to find the coordinate of the midpoint of each segment.

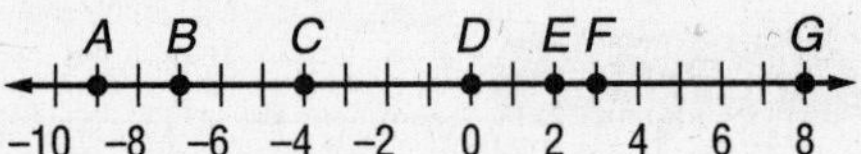

1. $\overline{CE}$

2. $\overline{DG}$

3. $\overline{AF}$

4. $\overline{EG}$

5. $\overline{AB}$

6. $\overline{BG}$

7. $\overline{BD}$

8. $\overline{DE}$

Find the coordinates of the midpoint of a segment having the given endpoints.

9. $A(0, 0)$, $B(12, 8)$

10. $R(-12, 8)$, $S(6, 12)$

11. $M(11, -2)$, $N(-9, 13)$

12. $E(-2, 6)$, $F(-9, 3)$

13. $S(10, -22)$, $T(9, 10)$

14. $M(-11, 2)$, $N(-19, 6)$

NAME ______________________________ DATE ____________ PERIOD _____

1-4 Study Guide and Intervention

Angle Measure

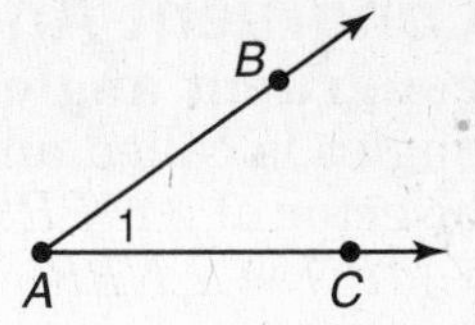

Measure Angles If two noncollinear **rays** have a common endpoint, they form an **angle**. The rays are the **sides** of the angle. The common endpoint is the **vertex**. The angle at the right can be named as $\angle A$, $\angle BAC$, $\angle CAB$, or $\angle 1$.

A **right angle** is an angle whose measure is 90. An **acute angle** has measure less than 90. An **obtuse angle** has measure greater than 90 but less than 180.

Example 1

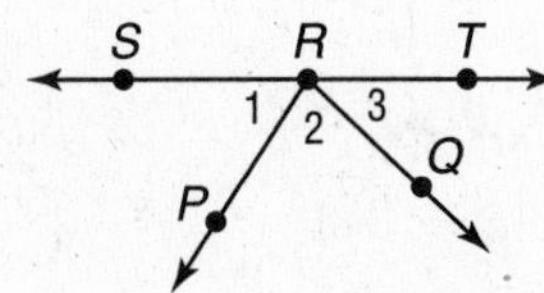

a. Name all angles that have *R* as a vertex.

Three angles are $\angle 1$, $\angle 2$, and $\angle 3$. For other angles, use three letters to name them: $\angle SRQ$, $\angle PRT$, and $\angle SRT$.

b. Name the sides of $\angle 1$.

$\overrightarrow{RS}$, $\overrightarrow{RP}$

Example 2 **Measure each angle and classify it as *right*, *acute*, or *obtuse*.**

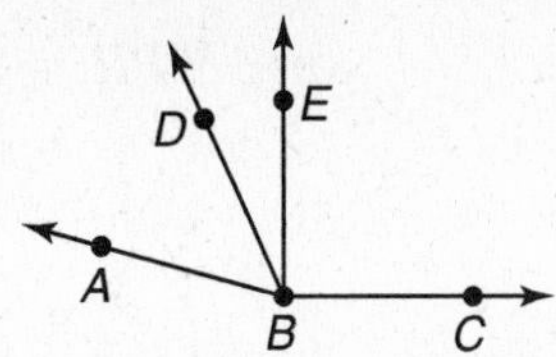

a. $\angle ABD$

Using a protractor, $m\angle ABD = 50$.

$50 < 90$, so $\angle ABD$ is an acute angle.

b. $\angle DBC$

Using a protractor, $m\angle DBC = 115$.

$180 > 115 > 90$, so $\angle DBC$ is an obtuse angle.

c. $\angle EBC$

Using a protractor, $m\angle EBC = 90$.

$\angle EBC$ is a right angle.

Exercises

Refer to the figure.

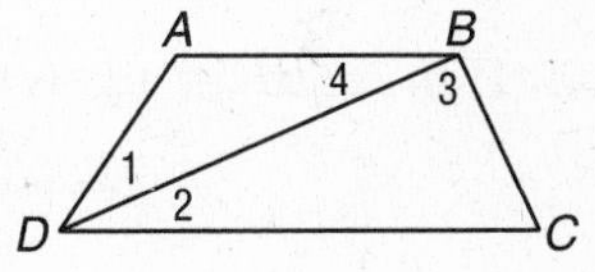

1. Name the vertex of $\angle 4$. B

2. Name the sides of $\angle BDC$. $\overrightarrow{BD}$ $\overrightarrow{DC}$

3. Write another name for $\angle DBC$. $\angle CBD$

Measure each angle in the figure and classify it as *right*, *acute*, or *obtuse*.

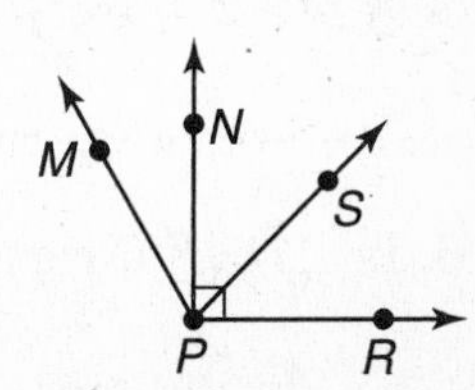

4. $\angle MPR$

5. $\angle RPN$

6. $\angle NPS$

NAME ______________________ DATE ____________ PERIOD _____

1-4 Study Guide and Intervention *(continued)*

Angle Measure

Congruent Angles Angles that have the same measure are **congruent angles**. A ray that divides an angle into two congruent angles is called an **angle bisector**. In the figure, $\overrightarrow{PN}$ is the angle bisector of $\angle MPR$. Point N lies in the interior of $\angle MPR$ and $\angle MPN \cong \angle NPR$.

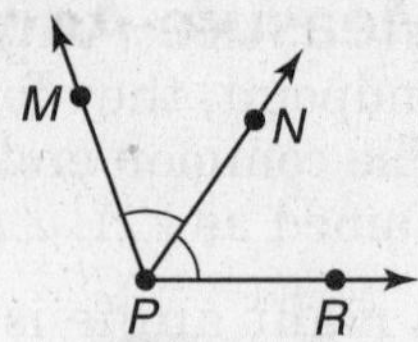

Example **Refer to the figure above. If $m\angle MPN = 2x + 14$ and $m\angle NPR = x + 34$, find x and find $m\angle MPR$.**

Since $\overrightarrow{PN}$ bisects $\angle MPR$, $\angle MPN \cong \angle NPR$, or $m\angle MPN = m\angle NPR$.

$$2x + 14 = x + 34$$
$$2x + 14 - x = x + 34 - x$$
$$x + 14 = 34$$
$$x + 14 - 14 = 34 - 14$$
$$x = 20$$

$$m\angle MPR = (2x + 14) + (x + 34)$$
$$= 54 + 54$$
$$= 108$$

Exercises

$\overrightarrow{QS}$ bisects $\angle PQT$, and $\overrightarrow{QP}$ and $\overrightarrow{QR}$ are opposite rays.

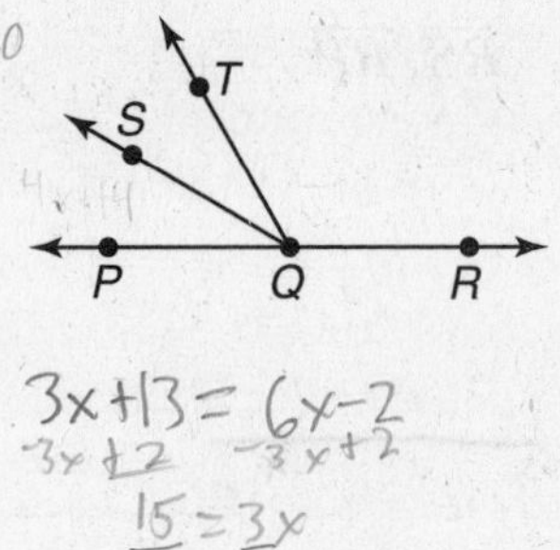

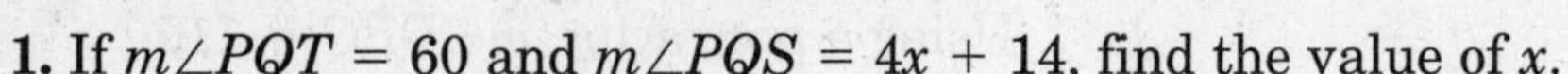

1. If $m\angle PQT = 60$ and $m\angle PQS = 4x + 14$, find the value of x.

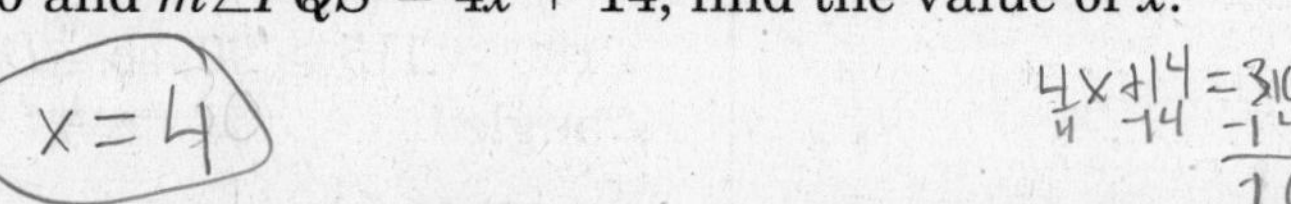

2. If $m\angle PQS = 3x + 13$ and $m\angle SQT = 6x - 2$, find $m\angle PQT$.

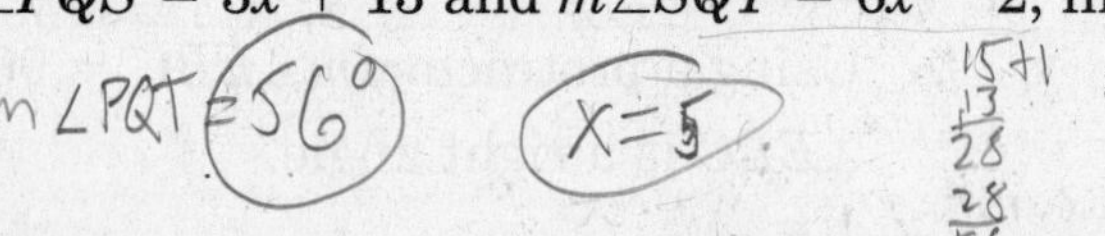

$\overrightarrow{BA}$ and $\overrightarrow{BC}$ are opposite rays, $\overrightarrow{BF}$ bisects $\angle CBE$, and $\overrightarrow{BD}$ bisects $\angle ABE$.

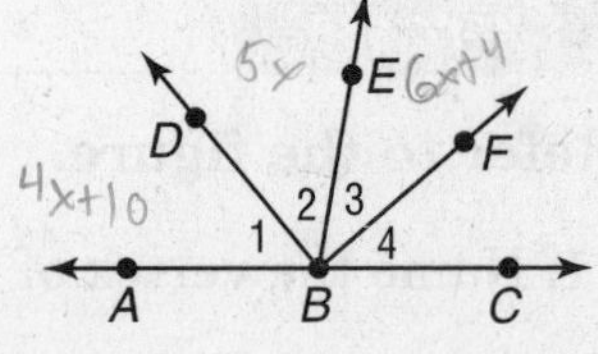

3. If $m\angle EBF = 6x + 4$ and $m\angle CBF = 7x - 2$, find $m\angle EBC$.

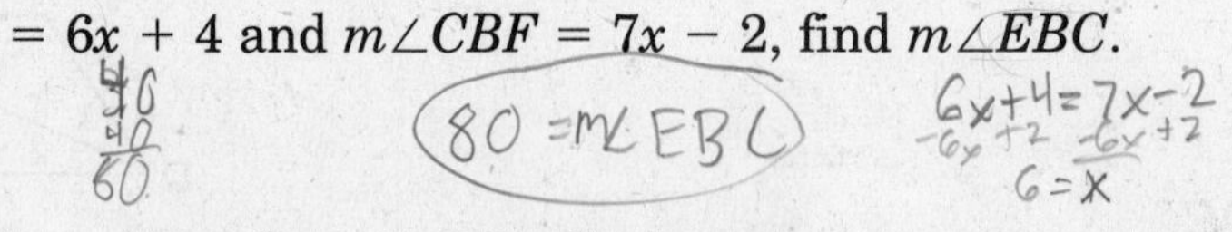

4. If $m\angle 1 = 4x + 10$ and $m\angle 2 = 5x$, find $m\angle 2$.

m∠2 = 50

5. If $m\angle 2 = 6y + 2$ and $m\angle 1 = 8y - 14$, find $m\angle ABE$.

m∠ABE=100

6. Is $\angle DBF$ a right angle? Explain.

Yes

1-5 Study Guide and Intervention

Angle Relationships

Pairs of Angles **Adjacent angles** are angles in the same plane that have a common vertex and a common side, but no common interior points. **Vertical angles** are two nonadjacent angles formed by two intersecting lines. A pair of adjacent angles whose noncommon sides are opposite rays is called a **linear pair**.

Example **Identify each pair of angles as *adjacent angles, vertical angles,* and/or as a *linear pair.***

a.

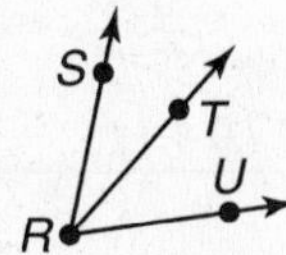

∠*SRT* and ∠*TRU* have a common vertex and a common side, but no common interior points. They are adjacent angles.

b.

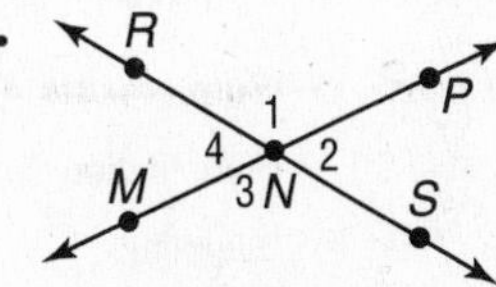

∠1 and ∠3 are nonadjacent angles formed by two intersecting lines. They are vertical angles. ∠2 and ∠4 are also vertical angles.

c.

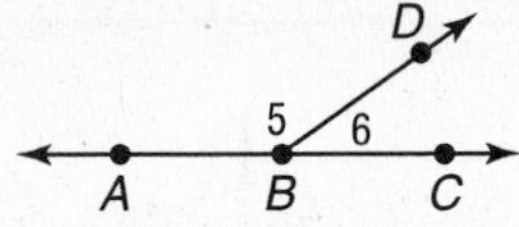

∠6 and ∠5 are adjacent angles whose noncommon sides are opposite rays. The angles form a linear pair.

d.

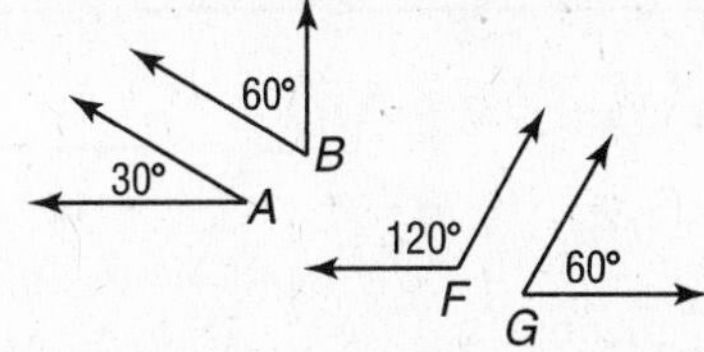

∠*A* and ∠*B* are two angles whose measures have a sum of 90. They are complementary. ∠*F* and ∠*G* are two angles whose measures have a sum of 180. They are supplementary.

Exercises

Identify each pair of angles as *adjacent, vertical,* and/or as a *linear pair.*

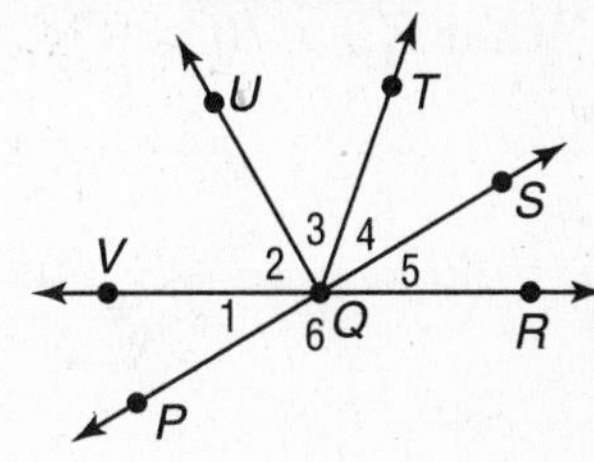

1. ∠1 and ∠2
adjacent

2. ∠1 and ∠6
linear pair

3. ∠1 and ∠5
vertical

4. ∠3 and ∠2
adjacent

For Exercises 5–7, refer to the figure at the right.

5. Identify two obtuse vertical angles. ∠SNU & ∠RNT

6. Identify two acute adjacent angles. ∠VNR ∠RNS

7. Identify an angle supplementary to ∠*TNU.* ∠TNR

8. Find the measures of two complementary angles if the difference in their measures is 18.

1-5 Study Guide and Intervention *(continued)*

Angle Relationships

Perpendicular Lines Lines, rays, and segments that form four right angles are **perpendicular**. The right angle symbol indicates that the lines are perpendicular. In the figure at the right, $\overleftrightarrow{AC}$ is perpendicular to $\overleftrightarrow{BD}$, or $\overleftrightarrow{AC} \perp \overleftrightarrow{BD}$.

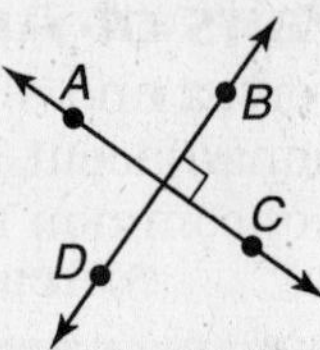

Example **Find x so that $\overline{DZ} \perp \overline{PZ}$.**

If $\overline{DZ} \perp \overline{PZ}$, then $m\angle DZP = 90$.

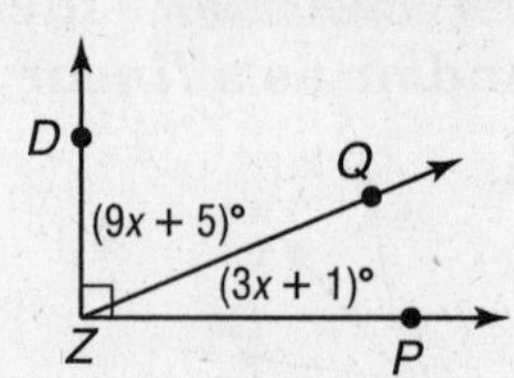

$m\angle DZQ + m\angle QZP = m\angle DZP$	Sum of parts = whole
$(9x + 5) + (3x + 1) = 90$	Substitution
$12x + 6 = 90$	Simplify.
$12x = 84$	Subtract 6 from each side.
$x = 7$	Divide each side by 12.

Exercises

1. Find x and y so that $\overleftrightarrow{NR} \perp \overleftrightarrow{MQ}$.

2. Find $m\angle MSN$.

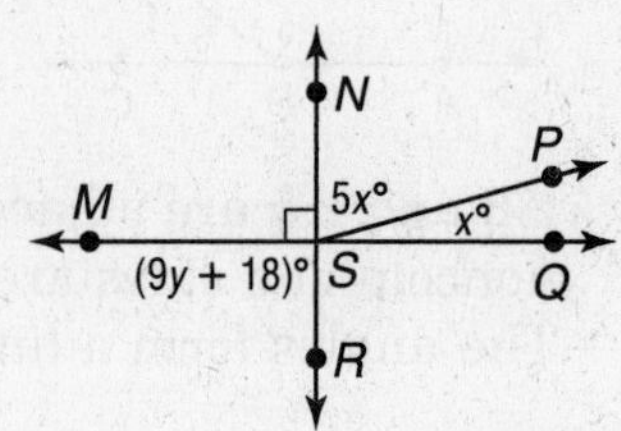

3. $m\angle EBF = 3x + 10$, $m\angle DBE = x$, and $\overrightarrow{BD} \perp \overrightarrow{BF}$. Find x.

4. If $m\angle EBF = 7y - 3$ and $m\angle FBC = 3y + 3$, find y so that $\overrightarrow{EB} \perp \overrightarrow{BC}$.

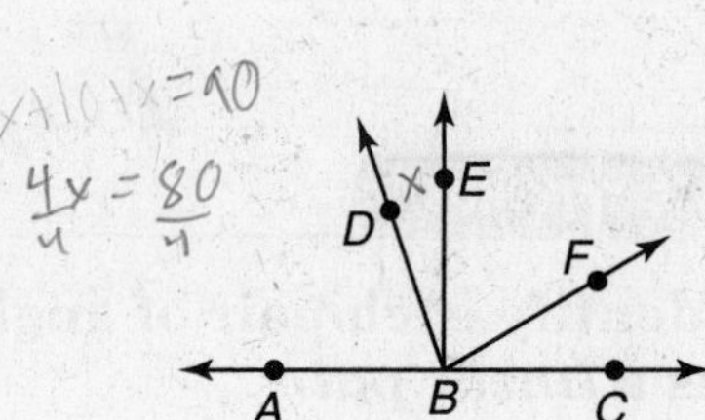

5. Find x, $m\angle PQS$, and $m\angle SQR$.

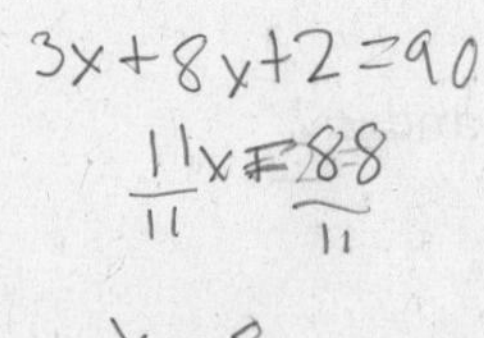

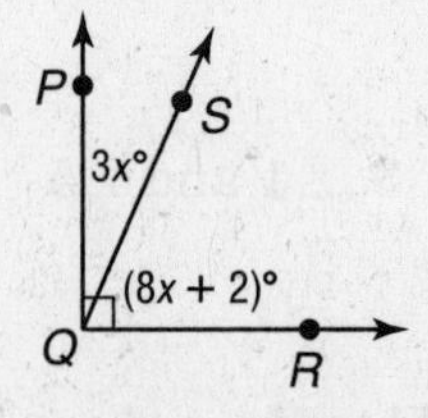

6. Find y, $m\angle RPT$, and $m\angle TPW$.

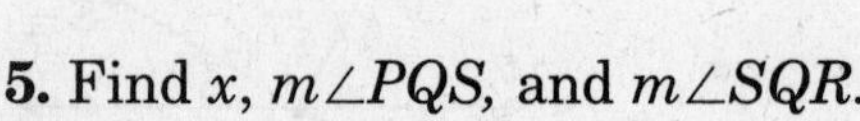

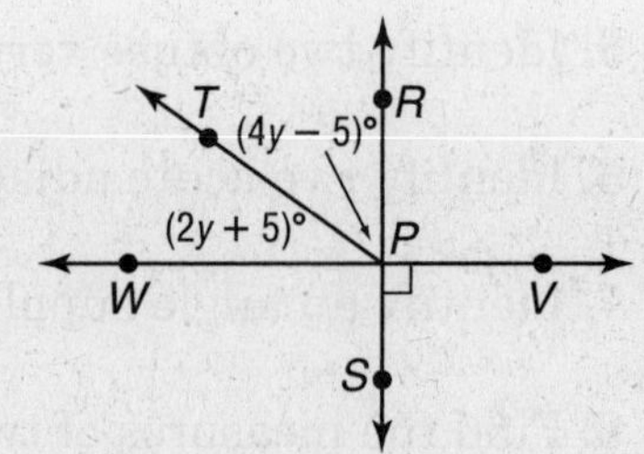

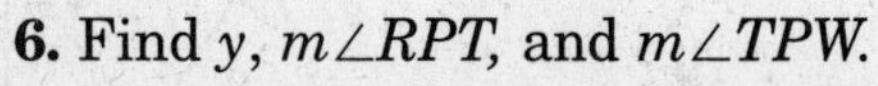

1-6 Study Guide and Intervention

Polygons

Polygons A **polygon** is a closed figure formed by a finite number of coplanar line segments. The sides that have a common endpoint must be noncollinear and each side intersects exactly two other sides at their endpoints. A polygon is named according to its number of sides. A **regular polygon** has congruent sides and congruent angles. A polygon can be **concave** or **convex**.

Example **Name each polygon by its number of sides. Then classify it as *concave* or *convex* and *regular* or *irregular*.**

a. 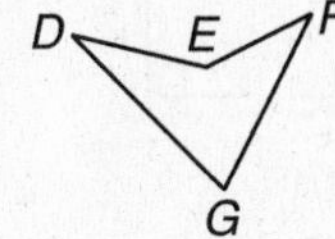

The polygon has 4 sides, so it is a quadrilateral. It is concave because part of $\overline{DE}$ or $\overline{EF}$ lies in the interior of the figure. Because it is concave, it cannot have all its angles congruent and so it is irregular.

b. 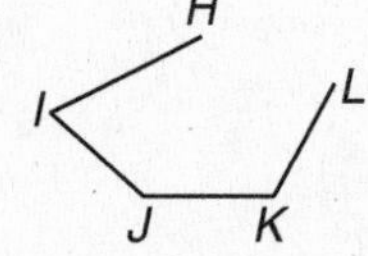

The figure is not closed, so it is not a polygon.

c. 

The polygon has 5 sides, so it is a pentagon. It is convex. All sides are congruent and all angles are congruent, so it is a regular pentagon.

d. 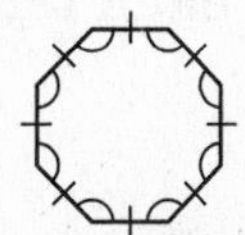

The figure has 8 congruent sides and 8 congruent angles. It is convex and is a regular octagon.

Exercises

Name each polygon by its number of sides. Then classify it as *concave* or *convex* and *regular* or *irregular*.

1. 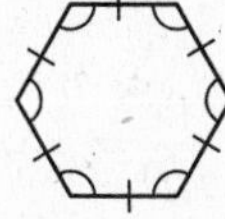

2. 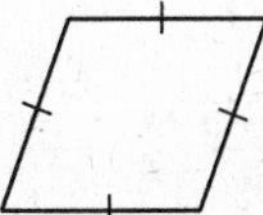

3. 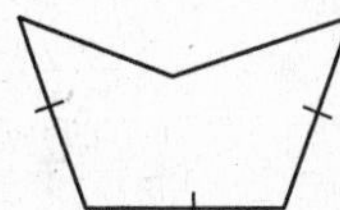

4. 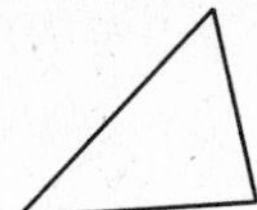

5. 

6. 

1-6 Study Guide and Intervention *(continued)*

Polygons

Perimeter The **perimeter** of a polygon is the sum of the lengths of all the sides of the polygon. There are special formulas for the perimeter of a square or a rectangle.

Example **Write an expression or formula for the perimeter of each polygon. Find the perimeter.**

a.

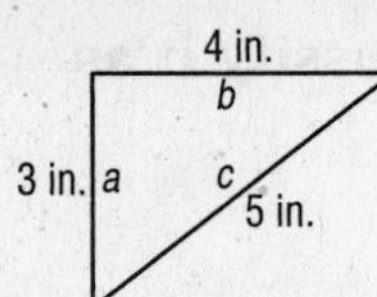

$P = a + b + c$
$= 3 + 4 + 5$
$= 12$ in.

b.

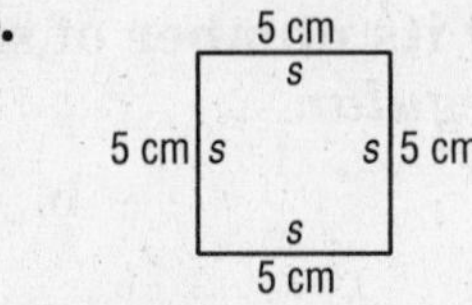

$P = 4s$
$= 4(5)$
$= 20$ cm

c.

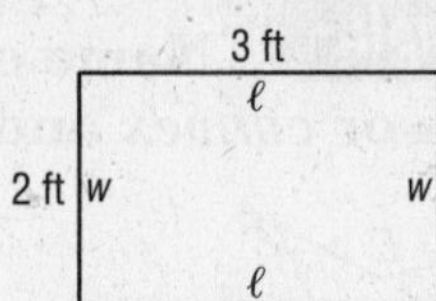

$P = 2\ell + 2w$
$= 2(3) + 2(2)$
$= 10$ ft

Exercises

Find the perimeter of each figure.

1.

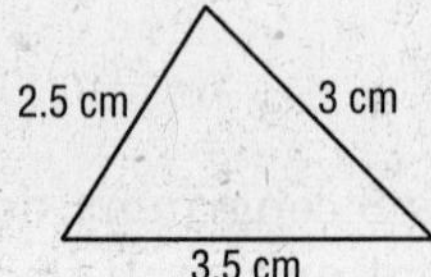

2.

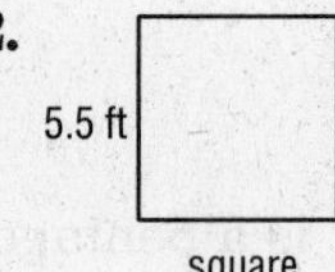

3.

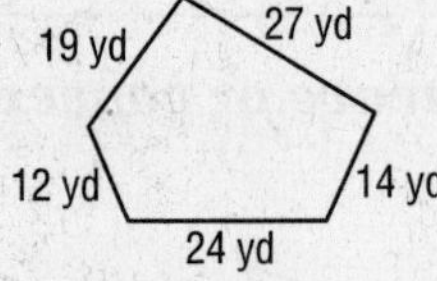

4.

Find the length of each side of the polygon for the given perimeter.

5. $P = 96$

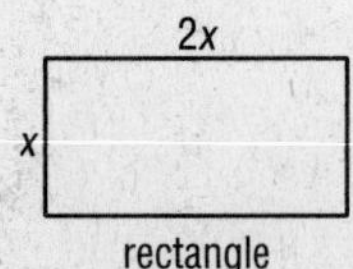

6. $P = 48$

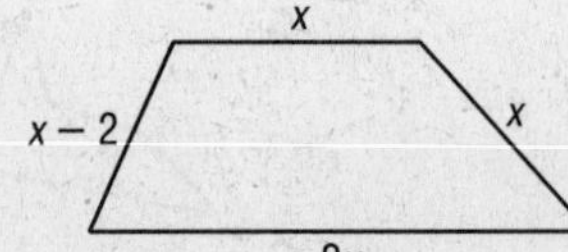

NAME ______________________ DATE __________ PERIOD _____

2-1 Study Guide and Intervention

Inductive Reasoning and Conjecture

Make Conjectures A **conjecture** is a guess based on analyzing information or observing a pattern. Making a conjecture after looking at several situations is called **inductive reasoning**.

Example 1 **Make a conjecture about the next number in the sequence 1, 3, 9, 27, 81.**

Analyze the numbers:
Notice that each number is a power of 3.

1	3	9	27	81
3^0	3^1	3^2	3^3	3^4

Conjecture: The next number will be 3^5 or 243.

Example 2 **Make a conjecture about the number of small squares in the next figure.**

Observe a pattern: The sides of the squares have measures 1, 2, and 3 units.
Conjecture: For the next figure, the side of the square will be 4 units, so the figure will have 16 small squares.

Exercises

Describe the pattern. Then make a conjecture about the next number in the sequence.

1. $-5, 10, -20, 40$

2. 1, 10, 100, 1000

3. $1, \frac{6}{5}, \frac{7}{5}, \frac{8}{5}$

Make a conjecture based on the given information. Draw a figure to illustrate your conjecture.

4. $A(-1, -1), B(2, 2), C(4, 4)$

5. $\angle 1$ and $\angle 2$ form a right angle.

6. $\angle ABC$ and $\angle DBE$ are vertical angles.

7. $\angle E$ and $\angle F$ are right angles.

2-1 Study Guide and Intervention *(continued)*

Inductive Reasoning and Conjecture

Find Counterexamples A conjecture is false if there is even one situation in which the conjecture is not true. The false example is called a **counterexample**.

Example **Determine whether the conjecture is *true* or *false*. If it is false, give a counterexample.**

Given: $\overline{AB} \cong \overline{BC}$

Conjecture: B is the midpoint of $\overline{AC}$.

Is it possible to draw a diagram with $\overline{AB} \cong \overline{BC}$ such that B is not the midpoint? This diagram is a counterexample because point B is not on $\overline{AC}$. The conjecture is false.

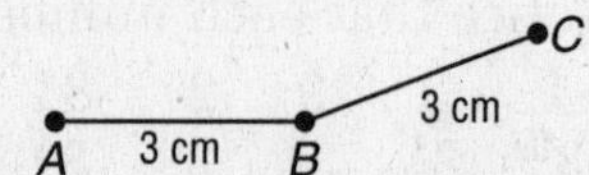

Exercises

Determine whether each conjecture is *true* or *false*. Give a counterexample for any false conjecture.

1. **Given:** Points A, B, and C are collinear.
   Conjecture: $AB + BC = AC$

2. **Given:** $\angle R$ and $\angle S$ are supplementary. $\angle R$ and $\angle T$ are supplementary.
   Conjecture: $\angle T$ and $\angle S$ are congruent.

3. **Given:** $\angle ABC$ and $\angle DEF$ are supplementary.
   Conjecture: $\angle ABC$ and $\angle DEF$ form a linear pair.

4. **Given:** $\overline{DE} \perp \overline{EF}$
   Conjecture: $\angle DEF$ is a right angle.

NAME ______________________________ DATE ____________ PERIOD _____

2-2 Study Guide and Intervention

Logic

Determine Truth Values A **statement** is any sentence that is either true or false. The truth or falsity of a statement is its **truth value**. A statement can be represented by using a letter. For example,

Statement p: Chicago is a city in Illinois. The truth value of statement p is true.

Several statements can be joined in a **compound statement**.

Statement p and statement q joined by the word *and* is a **conjunction**.	Statement p and statement q joined by the word *or* is a **disjunction**.	**Negation:** *not p* is the negation of the statement p.
Symbols: $p \wedge q$ (Read: *p and q*)	Symbols: $p \vee q$ (Read: *p or q*)	Symbols: $\sim p$ (Read: *not p*)
The conjunction $p \wedge q$ is true only when both p and q are true.	The disjunction $p \vee q$ is true if p is true, if q is true, or if both are true.	The statements p and $\sim p$ have opposite truth values.

Example 1 **Write a compound statement for each conjunction. Then find its truth value.**

p: An elephant is a mammal.

q: A square has four right angles.

a. $p \wedge q$

Join the statements with *and*: An elephant is a mammal and a square has four right angles. Both parts of the statement are true so the compound statement is true.

b. $\sim p \wedge q$

$\sim p$ is the statement "An elephant is not a mammal." Join $\sim p$ and q with the word *and*: An elephant is not a mammal and a square has four right angles. The first part of the compound statement, $\sim p$, is false. Therefore the compound statement is false.

Example 2 **Write a compound statement for each disjunction. Then find its truth value.**

p: A diameter of a circle is twice the radius.

q: A rectangle has four equal sides.

a. $p \vee q$

Join the statements p and q with the word *or*: A diameter of a circle is twice the radius or a rectangle has four equal sides. The first part of the compound statement, p, is true, so the compound statement is true.

b. $\sim p \vee q$

Join $\sim p$ and q with the word *or*: A diameter of a circle is not twice the radius or a rectangle has four equal sides. Neither part of the disjunction is true, so the compound statement is false.

Exercises

Write a compound statement for each conjunction and disjunction. Then find its truth value.

p: $10 + 8 = 18$ q: September has 30 days. r: A rectangle has four sides.

1. p and q

2. p or r

3. q or r

4. q and $\sim r$

NAME ______________________ DATE __________ PERIOD _____

2-2 Study Guide and Intervention *(continued)*

Logic

Truth Tables One way to organize the truth values of statements is in a **truth table**. The truth tables for negation, conjunction, and disjunction are shown at the right.

Negation	
p	~*p*
T	F
F	T

Conjunction		
p	*q*	$p \wedge q$
T	T	T
T	F	F
F	T	F
F	F	F

Disjunction		
p	*q*	$p \vee q$
T	T	T
T	F	T
F	T	T
F	F	F

Example 1 **Construct a truth table for the compound statement *q* or *r*. Use the disjunction table.**

q	*r*	*q* or *r*
T	T	T
T	F	T
F	T	T
F	F	F

Example 2 **Construct a truth table for the compound statement *p* and (*q* or *r*).**

Use the disjunction table for (*q* or *r*). Then use the conjunction table for *p* and (*q* or *r*).

p	*q*	*r*	*q* or *r*	*p* and (*q* or *r*)
T	T	T	T	T
T	T	F	T	T
T	F	T	T	T
T	F	F	F	F
F	T	T	T	F
F	T	F	T	F
F	F	T	T	F
F	F	F	F	F

Exercises

Contruct a truth table for each compound statement.

1. *p* or *r*

2. $\sim p \vee q$

3. $q \wedge \sim r$

4. $\sim p \wedge \sim r$

5. (*p* and *r*) or *q*

2-3 Study Guide and Intervention

Conditional Statements

If-then Statements An if-then statement is a statement such as "If you are reading this page, then you are studying math." A statement that can be written in if-then form is called a **conditional statement**. The phrase immediately following the word *if* is the **hypothesis**. The phrase immediately following the word *then* is the **conclusion**.

A conditional statement can be represented in symbols as $p \rightarrow q$, which is read "p implies q" or "if p, then q."

Example 1 **Identify the hypothesis and conclusion of the statement.**

If $\angle X \cong \angle R$ and $\angle R \cong \angle S$, then $\angle X \cong \angle S$.

hypothesis: $\angle X \cong \angle R$ and $\angle R \cong \angle S$; conclusion: $\angle X \cong \angle S$

Example 2 **Identify the hypothesis and conclusion. Write the statement in if-then form.**

You receive a free pizza with 12 coupons.

If you have 12 coupons, then you receive a free pizza.

hypothesis: you have 12 coupons; conclusion: you receive a free pizza

Exercises

Identify the hypothesis and conclusion of each statement.

1. If it is Saturday, then there is no school.

2. If $x - 8 = 32$, then $x = 40$.

3. If a polygon has four right angles, then the polygon is a rectangle.

Write each statement in if-then form.

4. All apes love bananas.

5. The sum of the measures of complementary angles is 90.

6. Collinear points lie on the same line.

Determine the truth value of the following statement for each set of conditions.
If it does not rain this Saturday, we will have a picnic.

7. It rains this Saturday, and we have a picnic.

8. It rains this Saturday, and we don't have a picnic.

9. It doesn't rain this Saturday, and we have a picnic.

10. It doesn't rain this Saturday, and we don't have a picnic.

Lesson 2-3

NAME ______________________________ DATE ____________ PERIOD _____

2-3 Study Guide and Intervention *(continued)*

Conditional Statements

Converse, Inverse, and Contrapositive If you change the hypothesis or conclusion of a conditional statement, you form a **related conditional**. This chart shows the three related conditionals, *converse*, *inverse*, and *contrapositive*, and how they are related to a conditional statement.

	Symbols	Formed by	Example
Conditional	$p \to q$	using the given hypothesis and conclusion	If two angles are vertical angles, then they are congruent.
Converse	$q \to p$	exchanging the hypothesis and conclusion	If two angles are congruent, then they are vertical angles.
Inverse	$\sim p \to \sim q$	replacing the hypothesis with its negation and replacing the conclusion with its negation	If two angles are not vertical angles, then they are not congruent.
Contrapositive	$\sim q \to \sim p$	negating the hypothesis, negating the conclusion, and switching them	If two angles are not congruent, then they are not vertical angles.

Just as a conditional statement can be true or false, the related conditionals also can be true or false. A conditional statement always has the same truth value as its contrapositive, and the converse and inverse always have the same truth value.

Exercises

Write the converse, inverse, and contrapositive of each conditional statement. Tell which statements are *true* and which statements are *false*.

1. If you live in San Diego, then you live in California.

2. If a polygon is a rectangle, then it is a square.

3. If two angles are complementary, then the sum of their measures is 90.

NAME ______________________________ DATE __________ PERIOD ____

2-4 Study Guide and Intervention

Deductive Reasoning

Law of Detachment **Deductive reasoning** is the process of using facts, rules, definitions, or properties to reach conclusions. One form of deductive reasoning that draws conclusions from a true conditional $p \to q$ and a true statement p is called the **Law of Detachment**.

Law of Detachment	If $p \to q$ is true and p is true, then q is true.
Symbols	$[(p \to q)] \land p] \to q$

Example **The statement *If two angles are supplementary to the same angle, then they are congruent* is a true conditional. Determine whether each conclusion is valid based on the given information. Explain your reasoning.**

a. Given: $\angle A$ and $\angle C$ are supplementary to $\angle B$.
Conclusion: $\angle A$ is congruent to $\angle C$.

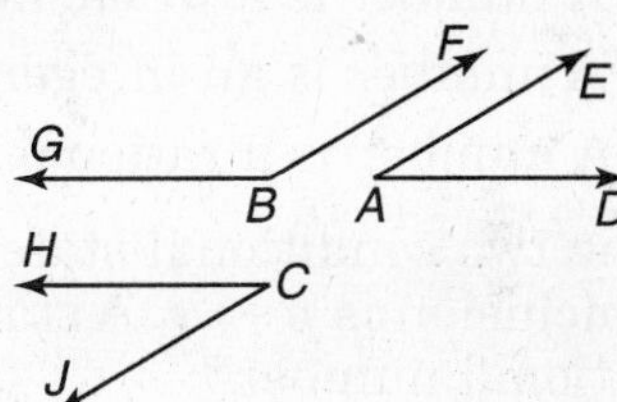

The statement *$\angle A$ and $\angle C$ are supplementary to $\angle B$* is the hypothesis of the conditional. Therefore, by the Law of Detachment, the conclusion is true.

b. Given: $\angle A$ is congruent to $\angle C$.
Conclusion: $\angle A$ and $\angle C$ are supplementary to $\angle B$.

The statement *$\angle A$ is congruent to $\angle C$* is not the hypothesis of the conditional, so the Law of Detachment cannot be used. The conclusion is not valid.

Exercises

Determine whether each conclusion is valid based on the true conditional given. If not, write *invalid*. Explain your reasoning.

If two angles are complementary to the same angle, then the angles are congruent.

1. Given: $\angle A$ and $\angle C$ are complementary to $\angle B$.
Conclusion: $\angle A$ is congruent to $\angle C$.

2. Given: $\angle A \cong \angle C$
Conclusion: $\angle A$ and $\angle C$ are complements of $\angle B$.

3. Given: $\angle E$ and $\angle F$ are complementary to $\angle G$.
Conclusion: $\angle E$ and $\angle F$ are vertical angles.

Lesson 2-4

2-4 Study Guide and Intervention *(continued)*

Deductive Reasoning

Law of Syllogism Another way to make a valid conclusion is to use the **Law of Syllogism**. It is similar to the Transitive Property.

Law of Syllogism	If $p \to q$ is true and $q \to r$ is true, then $p \to r$ is also true.
Symbols	$[(p \to q)] \wedge (q \to r)] \to (p \to r)$

Example **The two conditional statements below are true. Use the Law of Syllogism to find a valid conclusion. State the conclusion.**

(1) If a number is a whole number, then the number is an integer.
(2) If a number is an integer, then it is a rational number.

p: A number is a whole number.
q: A number is an integer.
r: A number is a rational number.

The two conditional statements are $p \to q$ and $q \to r$. Using the Law of Syllogism, a valid conclusion is $p \to r$. A statement of $p \to r$ is "if a number is a whole number, then it is a rational number."

Exercises

Determine whether you can use the Law of Syllogism to reach a valid conclusion from each set of statements.

1. If a dog eats Superdog Dog Food, he will be happy.
   Rover is happy.

2. If an angle is supplementary to an obtuse angle, then it is acute.
   If an angle is acute, then its measure is less than 90.

3. If the measure of $\angle A$ is less than 90, then $\angle A$ is acute.
   If $\angle A$ is acute, then $\angle A \cong \angle B$.

4. If an angle is a right angle, then the measure of the angle is 90.
   If two lines are perpendicular, then they form a right angle.

5. If you study for the test, then you will receive a high grade.
   Your grade on the test is high.

NAME ______________________________ DATE ______________ PERIOD _____

2-5 Study Guide and Intervention

Postulates and Paragraph Proofs

Points, Lines, and Planes In geometry, a **postulate** is a statement that is accepted as true. Postulates describe fundamental relationships in geometry.

Postulate: Through any two points, there is exactly one line.
Postulate: Through any three points not on the same line, there is exactly one plane.
Postulate: A line contains at least two points.
Postulate: A plane contains at least three points not on the same line.
Postulate: If two points lie in a plane, then the line containing those points lies in the plane.
Postulate: If two lines intersect, then their intersection is exactly one point.
Postulate: If two planes intersect, then their intersection is a line.

Example **Determine whether each statement is *always*, *sometimes*, or *never* true.**

a. There is exactly one plane that contains points *A*, *B*, and *C*.
Sometimes; if *A*, *B*, and *C* are collinear, they are contained in many planes. If they are noncollinear, then they are contained in exactly one plane.

b. Points *E* and *F* are contained in exactly one line.
Always; the first postulate states that there is exactly one line through any two points.

c. Two lines intersect in two distinct points *M* and *N*.
Never; the intersection of two lines is one point.

Exercises

Use postulates to determine whether each statement is *always*, *sometimes*, or *never* true.

1. A line contains exactly one point.

2. Noncollinear points *R*, *S*, and *T* are contained in exactly one plane.

3. Any two lines ℓ and m intersect.

4. If points *G* and *H* are contained in plane $\mathcal{M}$, then $\overline{GH}$ is perpendicular to plane $\mathcal{M}$.

5. Planes $\mathcal{R}$ and $\mathcal{S}$ intersect in point *T*.

6. If points *A*, *B*, and *C* are noncollinear, then segments $\overline{AB}$, $\overline{BC}$, and $\overline{CA}$ are contained in exactly one plane.

In the figure, $\overline{AC}$ and $\overline{DE}$ are in plane *Q* and $\overline{AC} \parallel \overline{DE}$. State the postulate that can be used to show each statement is true.

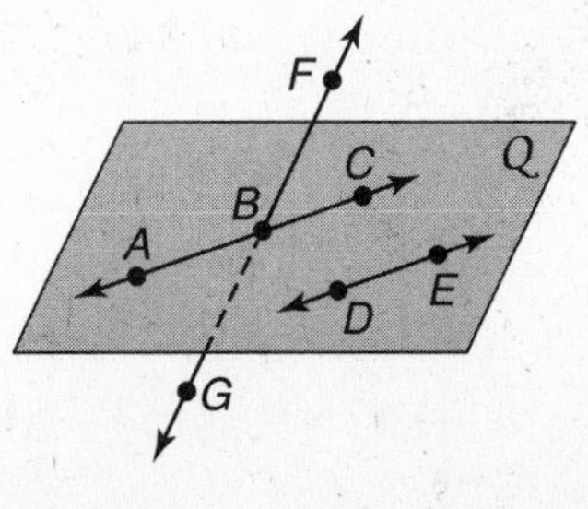

7. Exactly one plane contains points *F*, *B*, and *E*.

8. $\overleftrightarrow{BE}$ lies in plane *Q*.

Lesson 2-5

2-5 Study Guide and Intervention *(continued)*

Postulates and Paragraph Proofs

Paragraph Proofs A statement that can be proved true is called a **theorem**. You can use undefined terms, definitions, postulates, and already-proved theorems to prove other statements true.

A logical argument that uses deductive reasoning to reach a valid conclusion is called a **proof**. In one type of proof, a **paragraph proof**, you write a paragraph to explain why a statement is true.

Example **In $\triangle ABC$, $\overline{BD}$ is an angle bisector. Write a paragraph proof to show that $\angle ABD \cong \angle CBD$.**

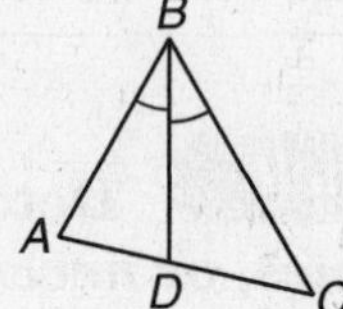

By definition, an angle bisector divides an angle into two congruent angles. Since $\overline{BD}$ is an angle bisector, $\angle ABC$ is divided into two congruent angles. Thus, $\angle ABD \cong \angle CBD$.

Exercises

1. Given that $\angle A \cong \angle D$ and $\angle D \cong \angle E$, write a paragraph proof to show that $\angle A \cong \angle E$.

2. It is given that $\overline{BC} \cong \overline{EF}$, M is the midpoint of $\overline{BC}$, and N is the midpoint of $\overline{EF}$. Write a paragraph proof to show that $BM = EN$.

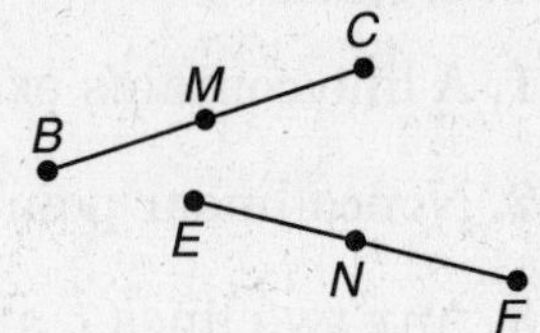

3. Given that S is the midpoint of $\overline{QP}$, T is the midpoint of $\overline{PR}$, and P is the midpoint of $\overline{ST}$, write a paragraph proof to show that $QS = TR$.

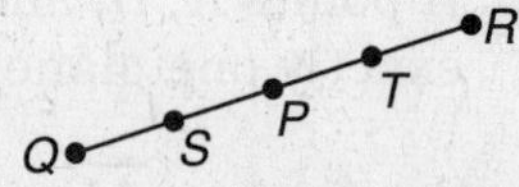

NAME ______________________ DATE ____________ PERIOD _____

2-6 Study Guide and Intervention

Algebraic Proof

Algebraic Proof The following properties of algebra can be used to justify the steps when solving an algebraic equation.

Property	Statement
Reflexive	For every number a, $a = a$.
Symmetric	For all numbers a and b, if $a = b$ then $b = a$.
Transitive	For all numbers a, b, and c, if $a = b$ and $b = c$ then $a = c$.
Addition and Subtraction	For all numbers a, b, and c, if $a = b$ then $a + c = b + c$ and $a - c = b - c$.
Multiplication and Division	For all numbers a, b, and c, if $a = b$ then $a \cdot c = b \cdot c$, and if $c \neq 0$ then $\frac{a}{c} = \frac{b}{c}$.
Substitution	For all numbers a and b, if $a = b$ then a may be replaced by b in any equation or expression.
Distributive	For all numbers a, b, and c, $a(b + c) = ab + ac$.

Example **Solve $6x + 2(x - 1) = 30$.**

Algebraic Steps	Properties
$6x + 2(x - 1) = 30$	Given
$6x + 2x - 2 = 30$	Distributive Property
$8x - 2 = 30$	Substitution
$8x - 2 + 2 = 30 + 2$	Addition Property
$8x = 32$	Substitution
$\frac{8x}{8} = \frac{32}{8}$	Division Property
$x = 4$	Substitution

Exercises

Complete each proof.

1. Given: $\frac{4x + 6}{2} = 9$
Prove: $x = 3$

Statements	Reasons
a. $\frac{4x + 6}{2} = 9$	a. given
b. $2\left(\frac{4x + 6}{2}\right) = 2(9)$	b. Mult. Prop.
c. $4x + 6 = 18$	c. Substitution
d. $4x + 6 - 6 = 18 - 6$	d. subtraction
e. $4x =$ 12	e. Substitution
f. $\frac{4x}{4} =$ $\frac{12}{4}$	f. Div. Prop.
g. x=3	g. Substitution

2. Given: $4x + 8 = x + 2$
Prove: $x = -2$

Statements	Reasons
a. $4x + 8 = x + 2$	a. given
b. $4x + 8 - x = x + 2 - x$	b. subtraction
c. $3x + 8 = 2$	c. Substitution
d. 3x+8-8=2-8	d. Subtr. Prop.
e. 3x=-6	e. Substitution
f. $\frac{3x}{3} = \frac{-6}{3}$	f. division
g. x=-2	g. Substitution

NAME ______________________ DATE __________ PERIOD _____

2-6 Study Guide and Intervention *(continued)*

Algebraic Proof

Geometric Proof Geometry deals with numbers as measures, so geometric proofs use properties of numbers. Here are some of the algebraic properties used in proofs.

Property	Segments	Angles
Reflexive	$AB = AB$	$m\angle A = m\angle A$
Symmetric	If $AB = CD$, then $CD = AB$.	If $m\angle A = m\angle B$, then $m\angle B = m\angle A$.
Transitive	If $AB = CD$ and $CD = EF$, then $AB = EF$.	If $m\angle 1 = m\angle 2$ and $m\angle 2 = m\angle 3$, then $m\angle 1 = m\angle 3$.

Example **Write a two-column proof.**

Given: $m\angle 1 = m\angle 2$, $m\angle 2 = m\angle 3$

Prove: $m\angle 1 = m\angle 3$

Proof:

Statements	Reasons
1. $m\angle 1 = m\angle 2$	1. Given
2. $m\angle 2 = m\angle 3$	2. Given
3. $m\angle 1 = m\angle 3$	3. Transitive Property

Exercises

State the property that justifies each statement.

1. If $m\angle 1 = m\angle 2$, then $m\angle 2 = m\angle 1$. Saimetrical
2. If $m\angle 1 = 90$ and $m\angle 2 = m\angle 1$, then $m\angle 2 = 90$. Substitution
3. If $AB = RS$ and $RS = WY$, then $AB = WY$. Transitive
4. If $AB = CD$, then $\frac{1}{2}AB = \frac{1}{2}CD$. Multiplication
5. If $m\angle 1 + m\angle 2 = 110$ and $m\angle 2 = m\angle 3$, then $m\angle 1 + m\angle 3 = 110$. Substitution
6. $RS = RS$ Reflexive
7. If $AB = RS$ and $TU = WY$, then $AB + TU = RS + WY$. Substitution
8. If $m\angle 1 = m\angle 2$ and $m\angle 2 = m\angle 3$, then $m\angle 1 = m\angle 3$. Transitive
9. A formula for the area of a triangle is $A = \frac{1}{2}bh$. Prove that bh is equal to 2 times the area of the triangle.

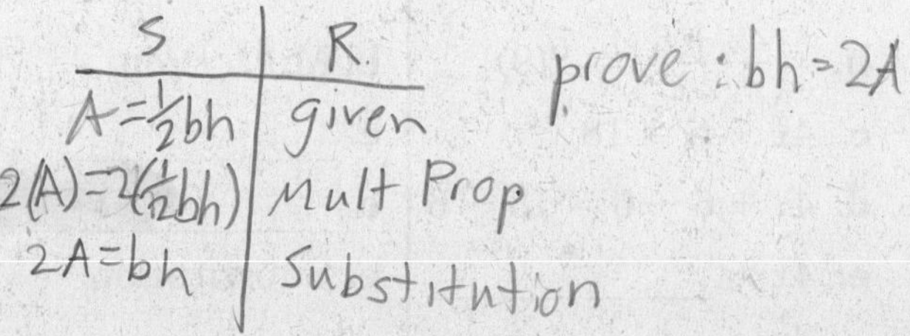

2-7 Study Guide and Intervention

Proving Segment Relationships

Segment Addition Two basic postulates for working with segments and lengths are the Ruler Postulate, which establishes number lines, and the Segment Addition Postulate, which describes what it means for one point to be between two other points.

Ruler Postulate	The points on any line or line segment can be paired with real numbers so that, given any two points *A* and *B* on a line, *A* corresponds to zero and *B* corresponds to a positive real number.
Segment Addition Postulate	*B* is between *A* and *C* if and only if $AB + BC = AC$.

Example **Write a two-column proof.**

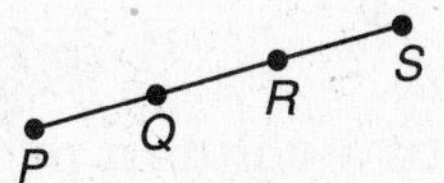

Given: Q is the midpoint of $\overline{PR}$.
R is the midpoint of $\overline{QS}$.

Prove: $PR = QS$

Statements	Reasons
1. Q is the midpoint of $\overline{PR}$.	**1.** Given
2. $PQ = QR$	**2.** Definition of midpoint
3. R is the midpoint of $\overline{QS}$.	**3.** Given
4. $QR = RS$	**4.** Definition of midpoint
5. $PQ + QR = QR + RS$	**5.** Addition Property
6. $PQ + QR = PR$, $QR + RS = QS$	**6.** Segment Addition Postulate
7. $PR = QS$	**7.** Substitution

Exercises

Complete each proof.

1. Given: $BC = DE$
Prove: $AB + DE = AC$

A B C D E

Statements	Reasons
a. $BC = DE$	**a.** ______________
b. ______________	**b.** Seg. Add. Post.
c. $AB + DE = AC$	**c.** ______________

2. Given: Q is between P and R, R is between Q and S, $PR = QS$.

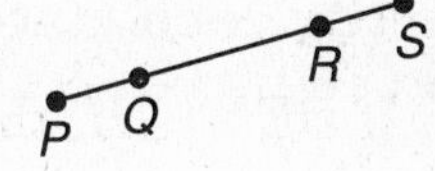

Prove: $PQ = RS$

Statements	Reasons
a. Q is between P and R.	**a.** Given
b. $PQ + QR = PR$	**b.** ______________
c. R is between Q and S.	**c.** ______________
d. ______________	**d.** Seg. Add. Post.
e. $PR = QS$	**e.** ______________
f. $PQ + QR = QR + RS$	**f.** ______________
g. $PQ + QR - QR = QR + RS - QR$	**g.** ______________
h. ______________	**h.** Substitution

NAME ______________________________ DATE ____________ PERIOD _____

2-7 Study Guide and Intervention *(continued)*

Proving Segment Relationships

Segment Congruence Three properties of algebra—the Reflexive, Symmetric, and Transitive Properties of Equality—have counterparts as properties of geometry. These properties can be proved as a theorem. As with other theorems, the properties can then be used to prove relationships among segments.

Segment Congruence Theorem	Congruence of segments is reflexive, symmetric, and transitive.
Reflexive Property	$\overline{AB} \cong \overline{AB}$
Symmetric Property	If $\overline{AB} \cong \overline{CD}$, then $\overline{CD} \cong \overline{AB}$.
Transitive Property	If $\overline{AB} \cong \overline{CD}$ and $\overline{CD} \cong \overline{EF}$, then $\overline{AB} \cong \overline{EF}$.

Example **Write a two-column proof.**

Given: $\overline{AB} \cong \overline{DE}$; $\overline{BC} \cong \overline{EF}$

Prove: $\overline{AC} \cong \overline{DF}$

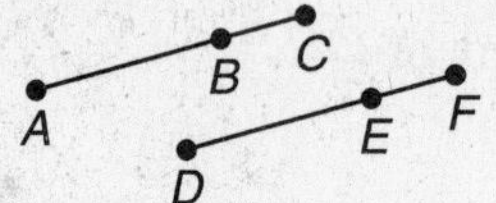

Statements	**Reasons**
1. $\overline{AB} \cong \overline{DE}$	**1.** Given
2. $AB = DE$	**2.** Definition of congruence of segments
3. $\overline{BC} \cong \overline{EF}$	**3.** Given
4. $BC = EF$	**4.** Definition of congruence of segments
5. $AB + BC = DE + EF$	**5.** Addition Property
6. $AB + BC = AC$, $DE + EF = DF$	**6.** Segment Addition Postulate
7. $AC = DF$	**7.** Substitution
8. $\overline{AC} \cong \overline{DF}$	**8.** Definition of congruence of segments

Exercises

Justify each statement with a property of congruence.

1. If $\overline{DE} \cong \overline{GH}$, then $\overline{GH} \cong \overline{DE}$.

2. If $\overline{AB} \cong \overline{RS}$ and $\overline{RS} \cong \overline{WY}$, then $\overline{AB} \cong \overline{WY}$.

3. $\overline{RS} \cong \overline{RS}$

4. Complete the proof.

Given: $\overline{PR} \cong \overline{QS}$

Prove: $\overline{PQ} \cong \overline{RS}$

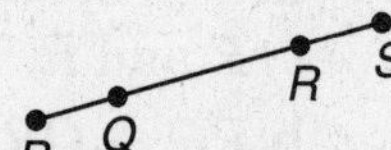

Statements	**Reasons**
a. $\overline{PR} \cong \overline{QS}$	**a.** ________
b. $PR = QS$	**b.** ________
c. $PQ + QR = PR$	**c.** ________
d. ________	**d.** Segment Addition Postulate
e. $PQ + QR = QR + RS$	**e.** ________
f. ________	**f.** Subtraction Property
g. ________	**g.** Definition of congruence of segments

2-8 Study Guide and Intervention

Proving Angle Relationships

Supplementary and Complementary Angles There are two basic postulates for working with angles. The Protractor Postulate assigns numbers to angle measures, and the Angle Addition Postulate relates parts of an angle to the whole angle.

Protractor Postulate	Given $\overrightarrow{AB}$ and a number r between 0 and 180, there is exactly one ray with endpoint A, extending on either side of $\overrightarrow{AB}$, such that the measure of the angle formed is r.
Angle Addition Postulate	R is in the interior of $\angle PQS$ if and only if $m\angle PQR + m\angle RQS = m\angle PQS$.

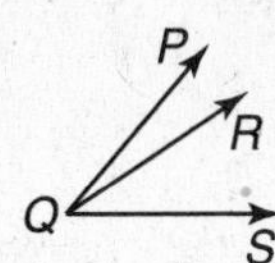

The two postulates can be used to prove the following two theorems.

Supplement Theorem Linear pair	If two angles form a linear pair, then they are supplementary angles. If $\angle 1$ and $\angle 2$ form a linear pair, then $m\angle 1 + m\angle 2 = 180$.
Complement Theorem	If the noncommon sides of two adjacent angles form a right angle, then the angles are complementary angles. If $\overleftrightarrow{GF} \perp \overleftrightarrow{GH}$, then $m\angle 3 + m\angle 4 = 90$.

Example 1 **If $\angle 1$ and $\angle 2$ form a linear pair and $m\angle 2 = 115$, find $m\angle 1$.**

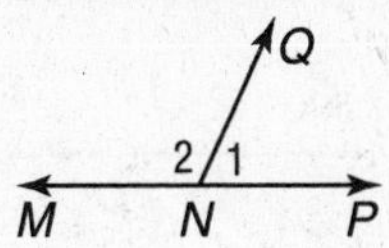

$m\angle 1 + m\angle 2 = 180$ Suppl. Theorem
$m\angle 1 + 115 = 180$ Substitution
$m\angle 1 = 65$ Subtraction Prop.

Example 2 **If $\angle 1$ and $\angle 2$ form a right angle and $m\angle 2 = 20$, find $m\angle 1$.**

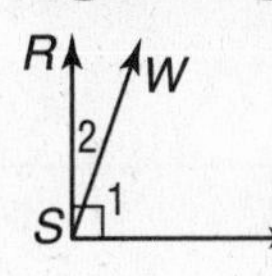

$m\angle 1 + m\angle 2 = 90$ Compl. Theorem
$m\angle 1 + 20 = 90$ Substitution
$m\angle 1 = 70$ Subtraction Prop.

Exercises

Find the measure of each numbered angle.

1.

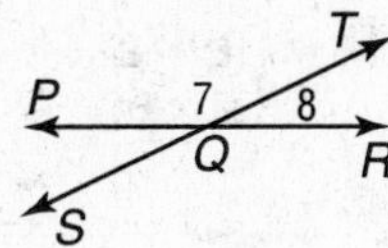

$m\angle 7 = 5x + 5,$
$m\angle 8 = x - 5$

2.

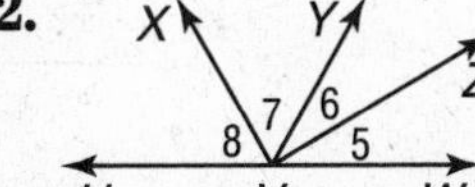

$m\angle 5 = 5x,\ m\angle 6 = 4x + 6,$
$m\angle 7 = 10x,$
$m\angle 8 = 12x - 12$

3.

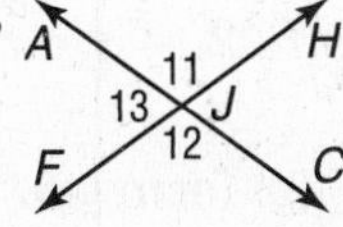

$m\angle 11 = 11x,$
$m\angle 12 = 10x + 10$

NAME ______________________ DATE ____________ PERIOD _____

2-8 Study Guide and Intervention *(continued)*

Proving Angle Relationships

Congruent and Right Angles Three properties of angles can be proved as theorems.

Congruence of angles is reflexive, symmetric, and transitive.	
Angles supplementary to the same angle or to congruent angles are congruent. 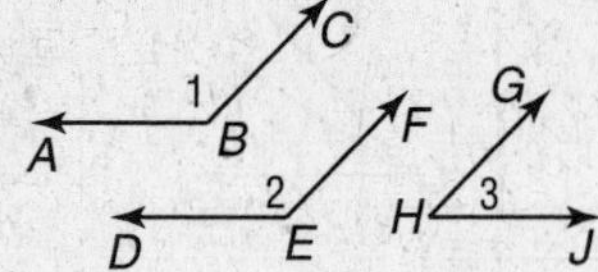 If ∠1 and ∠2 are supplementary to ∠3, then ∠1 ≅ ∠2.	Angles complementary to the same angle or to congruent angles are congruent. 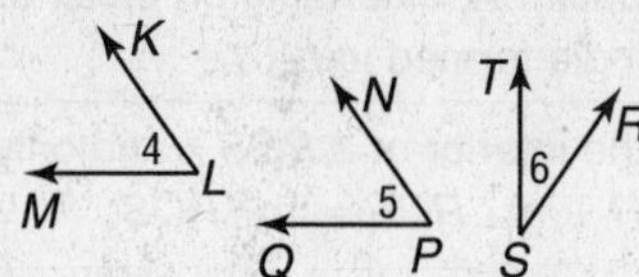 If ∠4 and ∠5 are complementary to ∠6, then ∠4 ≅ ∠5.

Example **Write a two-column proof.**

Given: ∠*ABC* and ∠*CBD* are complementary. ∠*DBE* and ∠*CBD* form a right angle.

Prove: ∠*ABC* ≅ ∠*DBE*

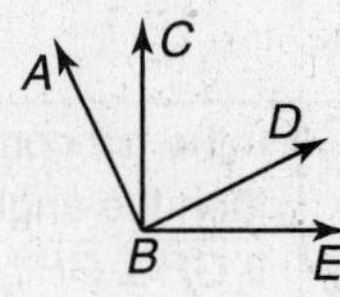

Statements	Reasons
1. ∠*ABC* and ∠*CBD* are complementary. ∠*DBE* and ∠*CBD* form a right angle.	**1.** Given
2. ∠*DBE* and ∠*CBD* are complementary.	**2.** Complement Theorem
3. ∠*ABC* ≅ ∠*DBE*	**3.** Angles complementary to the same ∠ are ≅.

Exercises

Complete each proof.

1. Given: $\overline{AB} \perp \overline{BC}$; ∠1 and ∠3 are complementary.

Prove: ∠2 ≅ ∠3

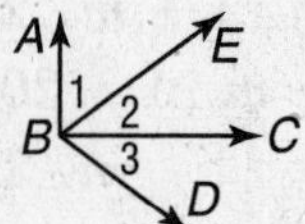

Statements	Reasons
a. $\overline{AB} \perp \overline{BC}$	**a.** given
b. ∠ABC = rt ∠	**b.** Definition of ⊥
c. m∠1 + m∠2 = m∠*ABC*	**c.** ∠ addition postulat
d. ∠1 and ∠2 form a rt ∠.	**d.** Substitution
e. ∠1 and ∠2 are compl.	**e.** def of complimentary
f. ∠1 and ∠3 are complimentary	**f.** Given
g. ∠2 ≅ ∠3	**g.** transitive

2. Given: ∠1 and ∠2 form a linear pair. m∠1 + m∠3 = 180

Prove: ∠2 ≅ ∠3

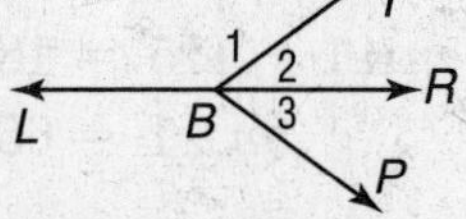

Statements	Reasons
a. ∠1 and ∠2 form a linear pair. m∠1 + m∠3 = 180	**a.** Given
b. m∠1 + m∠2 = 180	**b.** Suppl. Theorem
c. ∠1 is suppl. to ∠3.	**c.** def of suppl
d. ∠2 ≅ ∠3	**d.** ∠s suppl. to the same ∠ are ≅.

NAME ______________________ DATE ____________ PERIOD _____

3-1 Study Guide and Intervention

Parallel Lines and Transversals

Relationships Between Lines and Planes When two lines lie in the same plane and do not intersect, they are **parallel**. Lines that do not intersect and are not coplanar are **skew lines**. In the figure, ℓ is parallel to m, or $\ell \parallel m$. You can also write $\overleftrightarrow{PQ} \parallel \overleftrightarrow{RS}$. Similarly, if two planes do not intersect, they are **parallel planes**.

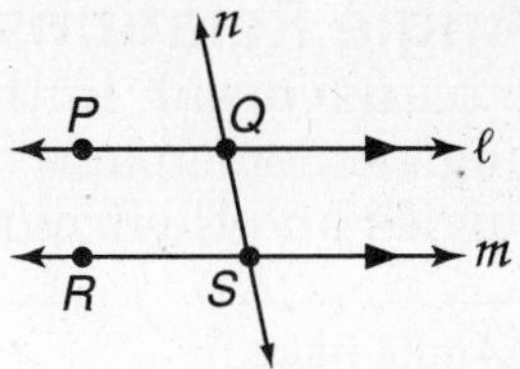

Example

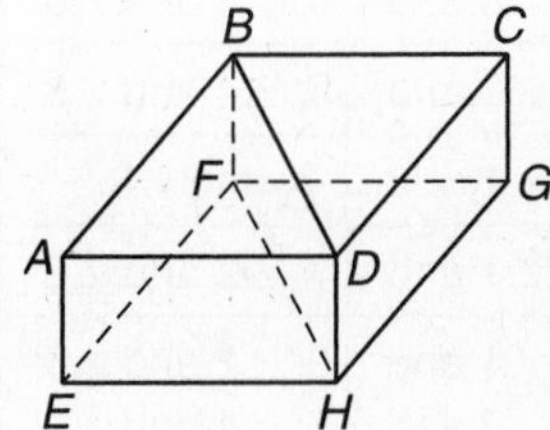

a. Name all planes that are parallel to plane *ABD*.
plane *EFH*

b. Name all segments that are parallel to $\overline{CG}$.
$\overline{BF}$, $\overline{DH}$, and $\overline{AE}$

c. Name all segments that are skew to $\overline{EH}$.
$\overline{BF}$, $\overline{CG}$, $\overline{BD}$, $\overline{CD}$, and $\overline{AB}$

Exercises

For Exercises 1–3, refer to the figure at the right.

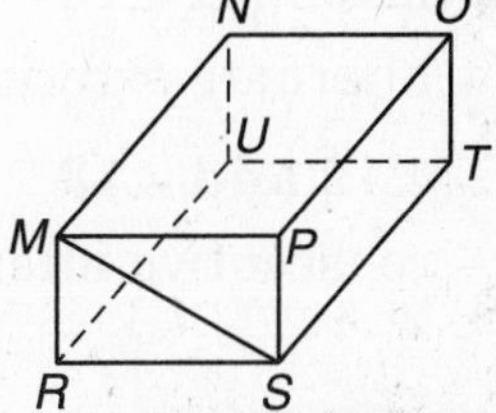

1. Name all planes that intersect plane *OPT*.

2. Name all segments that are parallel to $\overline{NU}$.

3. Name all segments that intersect $\overline{MP}$.

For Exercises 4–7, refer to the figure at the right.

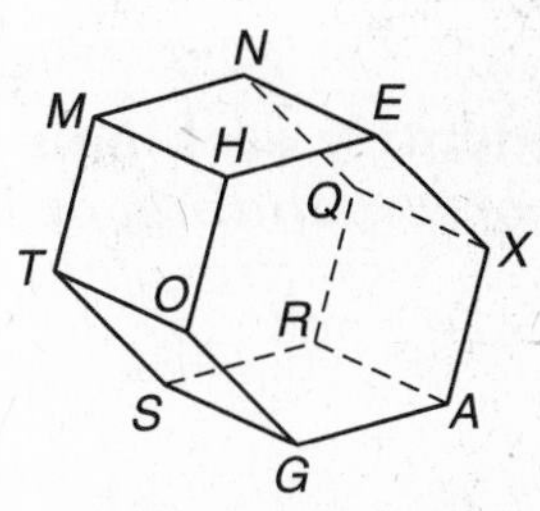

4. Name all segments parallel to $\overline{QX}$.

5. Name all planes that intersect plane *MHE*.

6. Name all segments parallel to $\overline{QR}$.

7. Name all segments skew to $\overline{AG}$.

Lesson 3-1

© Glencoe/McGraw-Hill

3-1 Study Guide and Intervention *(continued)*

Parallel Lines and Transversals

Angle Relationships A line that intersects two or more other lines in a plane is called a **transversal**. In the figure below, *t* is a transversal. Two lines and a transversal form eight angles. Some pairs of the angles have special names. The following chart lists the pairs of angles and their names.

Angle Pairs	Name
∠3, ∠4, ∠5, and ∠6	interior angles
∠3 and ∠5; ∠4 and ∠6	alternate interior angles
∠3 and ∠6; ∠4 and ∠5	consecutive interior angles
∠1, ∠2, ∠7, and ∠8	exterior angles
∠1 and ∠7; ∠2 and ∠8	alternate exterior angles
∠1 and ∠5; ∠2 and ∠6; ∠3 and ∠7; ∠4 and ∠8	corresponding angles

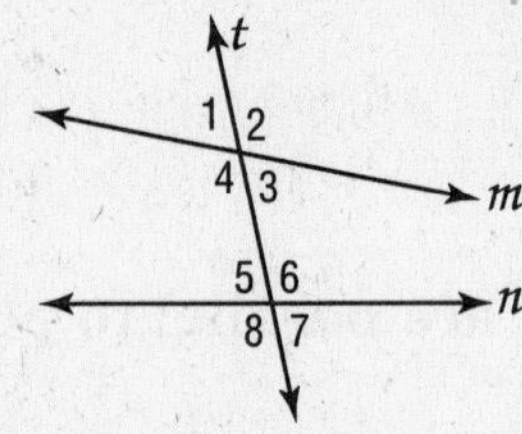

Example **Identify each pair of angles as *alternate interior, alternate exterior, corresponding,* or *consecutive interior* angles.**

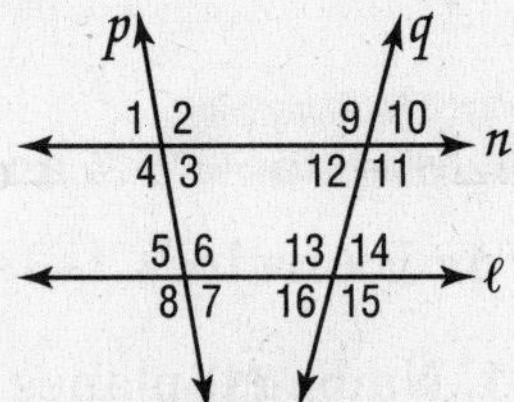

a. ∠10 and ∠16
alternate exterior angles

b. ∠4 and ∠12
corresponding angles

c. ∠12 and ∠13
consecutive interior angles

d. ∠3 and ∠9
alternate interior angles

Exercises

Use the figure in the Example for Exercises 1–12.

Name the transversal that forms each pair of angles.

1. ∠9 and ∠13

2. ∠5 and ∠14

3. ∠4 and ∠6

Identify each pair of angles as *alternate interior, alternate exterior, corresponding,* or *consecutive interior* angles.

4. ∠1 and ∠5

5. ∠6 and ∠14

6. ∠2 and ∠8

7. ∠3 and ∠11

8. ∠12 and ∠3

9. ∠4 and ∠6

10. ∠6 and ∠16

11. ∠11 and ∠14

12. ∠10 and ∠16

NAME ______________________ DATE __________ PERIOD _____

3-2 Study Guide and Intervention

Angles and Parallel Lines

Parallel Lines and Angle Pairs When two parallel lines are cut by a transversal, the following pairs of angles are congruent.

- corresponding angles
- alternate interior angles
- alternate exterior angles

Also, consecutive interior angles are supplementary.

Example **In the figure, $m\angle 2 = 75$. Find the measures of the remaining angles.**

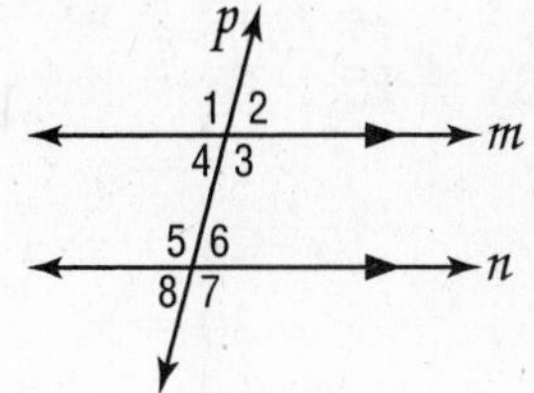

$m\angle 1 = 105$ $\angle 1$ and $\angle 2$ form a linear pair.
$m\angle 3 = 105$ $\angle 3$ and $\angle 2$ form a linear pair.
$m\angle 4 = 75$ $\angle 4$ and $\angle 2$ are vertical angles.
$m\angle 5 = 105$ $\angle 5$ and $\angle 3$ are alternate interior angles.
$m\angle 6 = 75$ $\angle 6$ and $\angle 2$ are corresponding angles.
$m\angle 7 = 105$ $\angle 7$ and $\angle 3$ are corresponding angles.
$m\angle 8 = 75$ $\angle 8$ and $\angle 6$ are vertical angles.

Exercises

In the figure, $m\angle 3 = 102$. Find the measure of each angle.

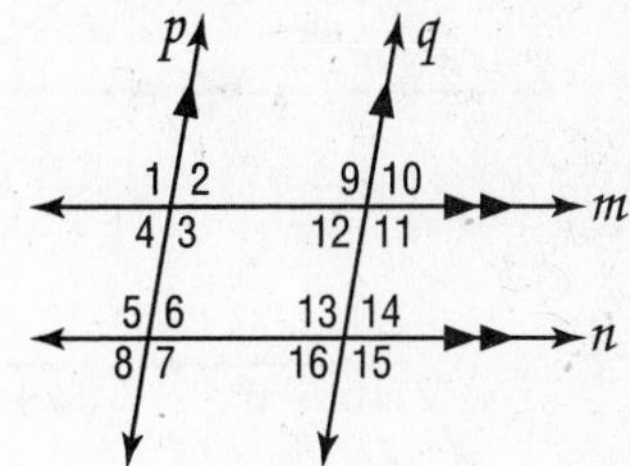

1. $\angle 5$ alternate interior 102°
2. $\angle 6$ 78° consec interior

3. $\angle 11$ 102
4. $\angle 7$ 102

5. $\angle 15$ 102
6. $\angle 14$ 78

In the figure, $m\angle 9 = 80$ and $m\angle 5 = 68$. Find the measure of each angle.

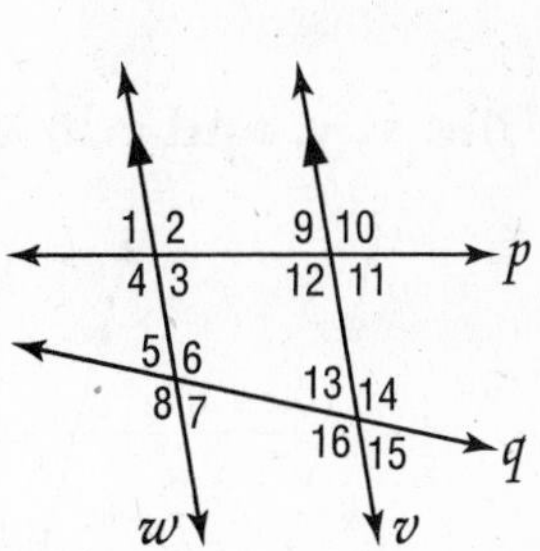

7. $\angle 12$
8. $\angle 1$

9. $\angle 4$
10. $\angle 3$

11. $\angle 7$
12. $\angle 16$

Lesson 3-2

NAME ______________________ DATE ____________ PERIOD _____

3-2 Study Guide and Intervention *(continued)*

Angles and Parallel Lines

Algebra and Angle Measures Algebra can be used to find unknown values in angles formed by a transversal and parallel lines.

Example **If $m\angle 1 = 3x + 15$, $m\angle 2 = 4x - 5$, $m\angle 3 = 5y$, and $m\angle 4 = 6z + 3$, find x and y.**

$p \parallel q$, so $m\angle 1 = m\angle 2$ because they are corresponding angles.

$$3x + 15 = 4x - 5$$
$$3x + 15 - 3x = 4x - 5 - 3x$$
$$15 = x - 5$$
$$15 + 5 = x - 5 + 5$$
$$20 = x$$

$r \parallel s$, so $m\angle 2 = m\angle 3$ because they are corresponding angles.

$$m\angle 2 = m\angle 3$$
$$75 = 5y$$
$$\frac{75}{5} = \frac{5y}{5}$$
$$15 = y$$

Exercises

Find x and y in each figure.

1.

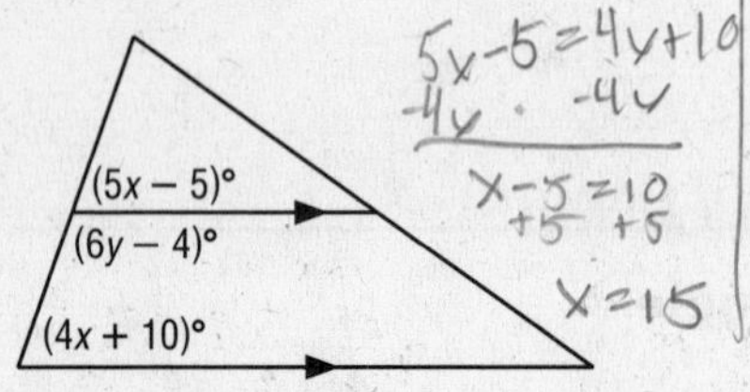

2.

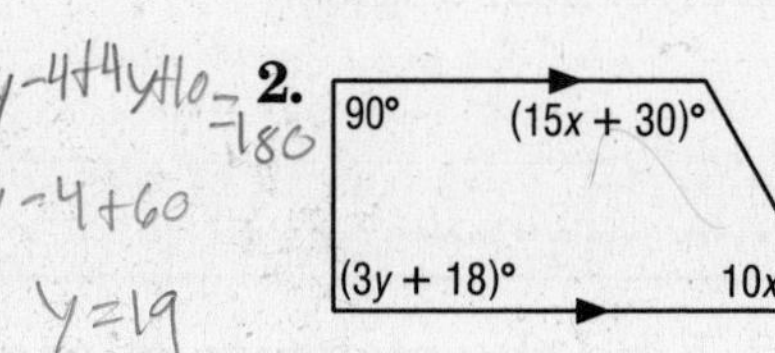

3.

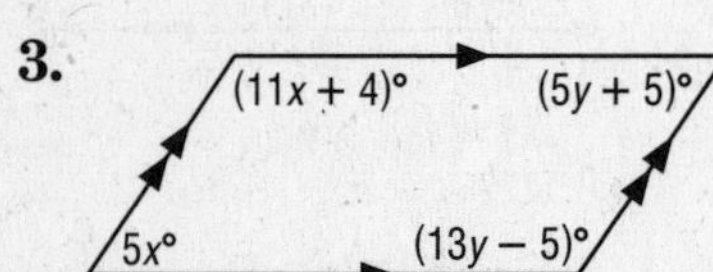

4.

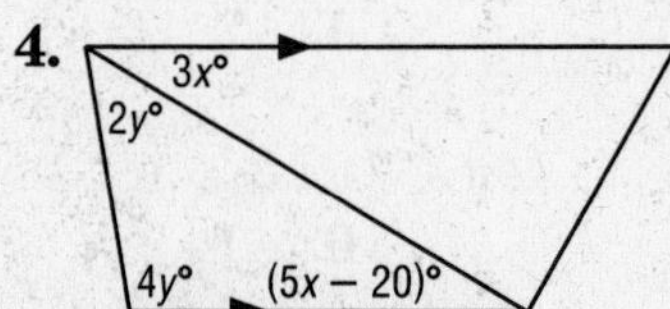

Find x, y, and z in each figure.

5.

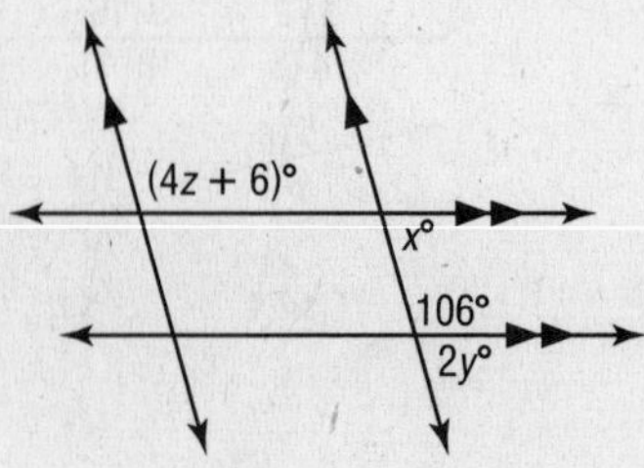

6.

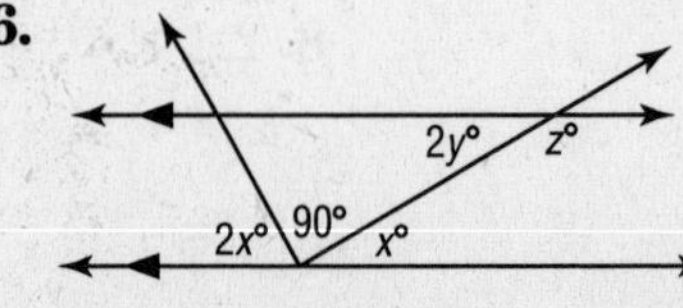

3-3 Study Guide and Intervention

Slopes of Lines

Slope of a Line The slope m of a line containing two points with coordinates (x_1, y_1) and (x_2, y_2) is given by the formula $m = \frac{y_2 - y_1}{x_2 - x_1}$, where $x_1 \neq x_2$.

Example **Find the slope of each line.**

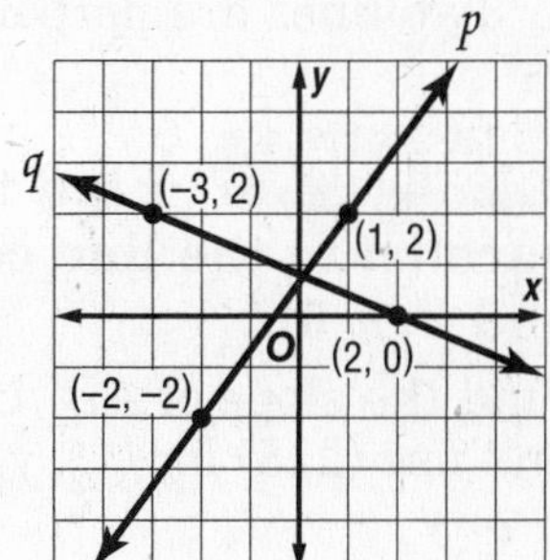

For line p, let (x_1, y_1) be (1, 2) and (x_2, y_2) be (−2, −2).

$$m = \frac{y_2 - y_1}{x_2 - x_1}$$
$$= \frac{-2 - 2}{-2 - 1} \text{ or } \frac{4}{3}$$

For line q, let (x_1, y_1) be (2, 0) and (x_2, y_2) be (−3, 2).

$$m = \frac{y_2 - y_1}{x_2 - x_1}$$
$$= \frac{2 - 0}{-3 - 2} \text{ or } -\frac{2}{5}$$

Exercises

Determine the slope of the line that contains the given points.

1. $J(0, 0)$, $K(-2, 8)$

2. $R(-2, -3)$, $S(3, -5)$

3. $L(1, -2)$, $N(-6, 3)$

4. $P(-1, 2)$, $Q(-9, 6)$

5. $T(1, -2)$, $U(6, -2)$

6. $V(-2, 10)$, $W(-4, -3)$

Find the slope of each line.

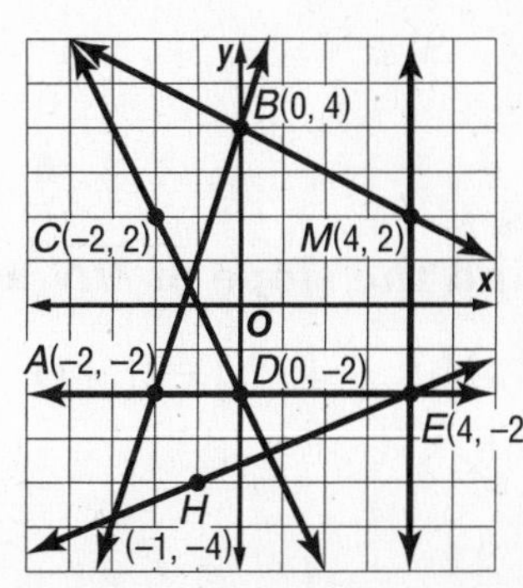

7. $\overleftrightarrow{AB}$

8. $\overleftrightarrow{CD}$

9. $\overleftrightarrow{EM}$

10. $\overleftrightarrow{AE}$

11. $\overleftrightarrow{EH}$

12. $\overleftrightarrow{BM}$

Lesson 3-3

3-3 Study Guide and Intervention *(continued)*

Slopes of Lines

Parallel and Perpendicular Lines If you examine the slopes of pairs of parallel lines and the slopes of pairs of perpendicular lines, where neither line in each pair is vertical, you will discover the following properties.

Two lines have the same slope if and only if they are parallel.

Two lines are perpendicular if and only if the product of their slopes is -1.

Example 1 **Find the slope of a line parallel to the line containing $A(-3, 4)$ and $B(2, 5)$.**

Find the slope of $\overleftrightarrow{AB}$. Use $(-3, 4)$ for (x_1, y_1) and use $(2, 5)$ for (x_2, y_2).

$$m = \frac{y_2 - y_1}{x_2 - x_1}$$

$$= \frac{5 - 4}{2 - (-3)} \text{ or } \frac{1}{5}$$

The slope of any line parallel to $\overleftrightarrow{AB}$ must be $\frac{1}{5}$.

Example 2 **Find the slope of a line perpendicular to $\overleftrightarrow{PQ}$ for $P(-2, -4)$ and $Q(4, 3)$.**

Find the slope of $\overleftrightarrow{PQ}$. Use $(-2, -4)$ for (x_1, y_1) and use $(4, 3)$ for (x_2, y_2).

$$m = \frac{y_2 - y_1}{x_2 - x_1}$$

$$= \frac{3 - (-4)}{4 - (-2)} \text{ or } \frac{7}{6}$$

Since $\frac{7}{6} \cdot \left(-\frac{6}{7}\right) = -1$, the slope of any line perpendicular to $\overleftrightarrow{PQ}$ must be $-\frac{6}{7}$.

Exercises

Determine whether $\overleftrightarrow{MN}$ and $\overleftrightarrow{RS}$ are *parallel, perpendicular*, or *neither*.

1. $M(0, 3)$, $N(2, 4)$, $R(2, 1)$, $S(8, 4)$

2. $M(-1, 3)$, $N(0, 5)$, $R(2, 1)$, $S(6, -1)$

3. $M(-1, 3)$, $N(4, 4)$, $R(3, 1)$, $S(-2, 2)$

4. $M(0, -3)$, $N(-2, -7)$, $R(2, 1)$, $S(0, -3)$

5. $M(-2, 2)$, $N(1, -3)$, $R(-2, 1)$, $S(3, 4)$

6. $M(0, 0)$, $N(2, 4)$, $R(2, 1)$, $S(8, 4)$

Find the slope of $\overleftrightarrow{MN}$ and the slope of any line perpendicular to $\overleftrightarrow{MN}$.

7. $M(2, -4)$, $N(-2, -1)$

8. $M(1, 3)$, $N(-1, 5)$

9. $M(4, -2)$, $N(5, 3)$

10. $M(2, -3)$, $N(-4, 1)$

NAME ______________________ DATE __________ PERIOD _____

3-4 Study Guide and Intervention

Equations of Lines

Write Equations of Lines You can write an equation of a line if you are given any of the following:

- the slope and the y-intercept,
- the slope and the coordinates of a point on the line, or
- the coordinates of two points on the line.

If m is the slope of a line, b is its y-intercept, and (x_1, y_1) is a point on the line, then:

- the **slope-intercept form** of the equation is $y = mx + b$,
- the **point-slope form** of the equation is $y - y_1 = m(x - x_1)$.

Example 1 **Write an equation in slope-intercept form of the line with slope −2 and y-intercept 4.**

$y = mx + b$ Slope-intercept form

$y = -2x + 4$ $m = -2$, $b = 4$

The slope-intercept form of the equation of the line is $y = -2x + 4$.

Example 2 **Write an equation in point-slope form of the line with slope $-\frac{3}{4}$ that contains (8, 1).**

$y - y_1 = m(x - x_1)$ Point-slope form

$y - 1 = -\frac{3}{4}(x - 8)$ $m = -\frac{3}{4}$, $(x_1, y_1) = (8, 1)$

The point-slope form of the equation of the line is $y - 1 = -\frac{3}{4}(x - 8)$.

Exercises

Write an equation in slope-intercept form of the line having the given slope and y-intercept.

1. m: 2, y-intercept: -3

2. m: $-\frac{1}{2}$, y-intercept: 4

3. m: $\frac{1}{4}$, y-intercept: 5

4. m: 0, y-intercept: -2

5. m: $-\frac{5}{3}$, y-intercept: $\frac{1}{3}$

6. m: -3, y-intercept: -8

Write an equation in point-slope form of the line having the given slope that contains the given point.

7. $m = \frac{1}{2}$, $(3, -1)$

8. $m = -2$, $(4, -2)$

9. $m = -1$, $(-1, 3)$

10. $m = \frac{1}{4}$, $(-3, -2)$

11. $m = -\frac{5}{2}$, $(0, -3)$

12. $m = 0$, $(-2, 5)$

Lesson 3-4

3-4 Study Guide and Intervention *(continued)*

Equations of Lines

Write Equations to Solve Problems Many real-world situations can be modeled using linear equations.

Example **Donna offers computer services to small companies in her city. She charges $55 per month for maintaining a web site and $45 per hour for each service call.**

a. Write an equation to represent the total monthly cost C for maintaining a web site and for h hours of service calls.

For each hour, the cost increases $45. So the rate of change, or slope, is 45. The y-intercept is located where there are 0 hours, or $55.

$C = mh + b$

$= 45h + 55$

b. Donna may change her costs to represent them by the equation $C = 25h + 125$, where $125 is the fixed monthly fee for a web site and the cost per hour is $25. Compare her new plan to the old one if a company has $5\frac{1}{2}$ hours of service calls. Under which plan would Donna earn more?

First plan

For $5\frac{1}{2}$ hours of service Donna would earn

$C = 45h + 55 = 45\left(5\frac{1}{2}\right) + 55$

$= 247.5 + 55$ or $302.50

Second Plan

For $5\frac{1}{2}$ hours of service Donna would earn

$C = 25h + 125 = 25(5.5) + 125$

$= 137.5 + 125$ or $262.50

Donna would earn more with the first plan.

Exercises

For Exercises 1–4, use the following information.

Jerri's current satellite television service charges a flat rate of $34.95 per month for the basic channels and an additional $10 per month for each premium channel. A competing satellite television service charges a flat rate of $39.99 per month for the basic channels and an additional $8 per month for each premium channel.

1. Write an equation in slope-intercept form that models the total monthly cost for each satellite service, where p is the number of premium channels.

2. If Jerri wants to include three premium channels in her package, which service would be less, her current service or the competing service?

3. A third satellite company charges a flat rate of $69 for all channels, including the premium channels. If Jerri wants to add a fourth premium channel, which service would be least expensive?

4. Write a description of how the fee for the number of premium channels is reflected in the equation.

NAME ______________________ DATE __________ PERIOD _____

3-5 Study Guide and Intervention

Proving Lines Parallel

Identify Parallel Lines If two lines in a plane are cut by a transversal and certain conditions are met, then the lines must be parallel.

If	then
• corresponding angles are congruent, • alternate exterior angles are congruent, • consecutive interior angles are supplementary, • alternate interior angles are congruent, or • two lines are perpendicular to the same line,	the lines are parallel.

Example 1 **If $m\angle 1 = m\angle 2$, determine which lines, if any, are parallel.**

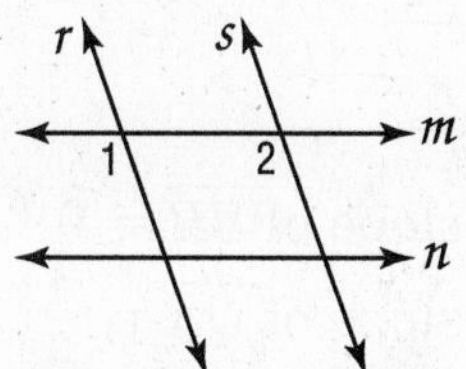

Since $m\angle 1 = m\angle 2$, then $\angle 1 \cong \angle 2$. $\angle 1$ and $\angle 2$ are congruent corresponding angles, so $r \parallel s$.

Example 2 **Find x and $m\angle ABC$ so that $m \parallel n$.**

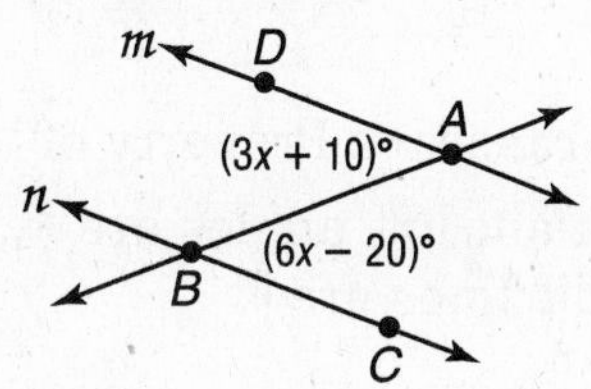

We can conclude that $m \parallel n$ if alternate interior angles are congruent.

$$m\angle BAD = m\angle ABC$$
$$3x + 10 = 6x - 20$$
$$10 = 3x - 20$$
$$30 = 3x$$
$$10 = x$$

$$m\angle ABC = 6x - 20$$
$$= 6(10) - 20 \text{ or } 40$$

Exercises

Find x so that $\ell \parallel m$.

1.

2.
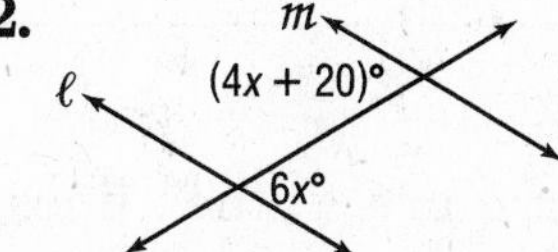

3.
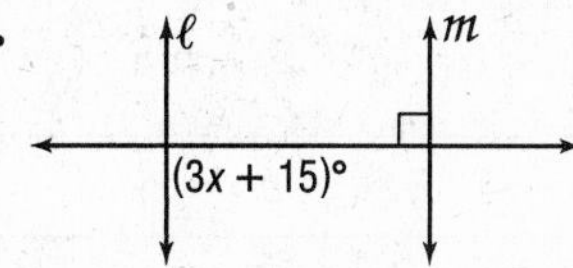

4.
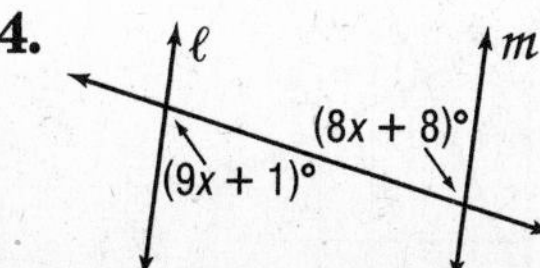

5.

6.
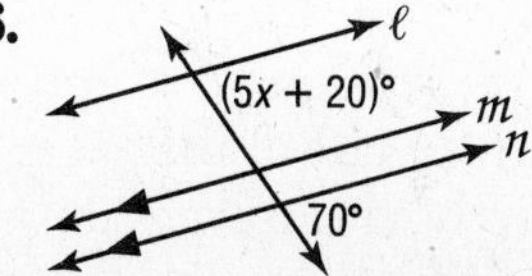

NAME ______________________ DATE __________ PERIOD ____

3-5 Study Guide and Intervention *(continued)*

Proving Lines Parallel

Prove Lines Parallel You can prove that lines are parallel by using postulates and theorems about pairs of angles. You also can use slopes of lines to prove that two lines are parallel or perpendicular.

Example

a. Given: $\angle 1 \cong \angle 2$, $\angle 1 \cong \angle 3$

Prove: $\overline{AB} \parallel \overline{DC}$

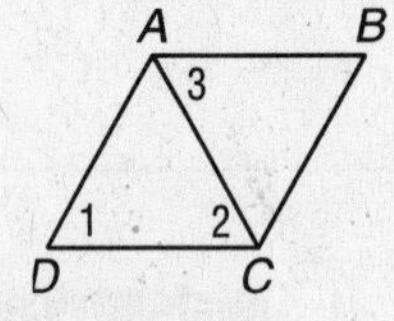

Statements	Reasons
1. $\angle 1 \cong \angle 2$ $\angle 1 \cong \angle 3$	1. Given
2. $\angle 2 \cong \angle 3$	2. Transitive Property of $\cong$
3. $\overline{AB} \parallel \overline{DC}$	3. If alt. int. angles are $\cong$, then the lines are $\parallel$.

b. Which lines are parallel? Which lines are perpendicular?

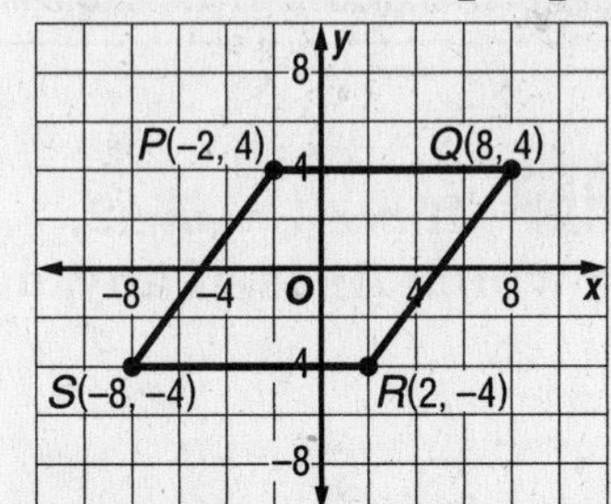

slope of $\overline{PQ} = 0$ slope of $\overline{SR} = 0$

slope of $\overline{PS} = \frac{4}{3}$ slope of $\overline{QR} = \frac{4}{3}$

slope of $\overline{PR} = -2$ slope of $\overline{SQ} = \frac{1}{2}$

So $\overline{PQ} \parallel \overline{SR}$, $\overline{PS} \parallel \overline{QR}$, and $\overline{PR} \perp \overline{SQ}$.

Exercises

For Exercises 1–6, fill in the blanks.

Given: $\angle 1 \cong \angle 5$, $\angle 15 \cong \angle 5$

Prove: $\ell \parallel m$, $r \parallel s$

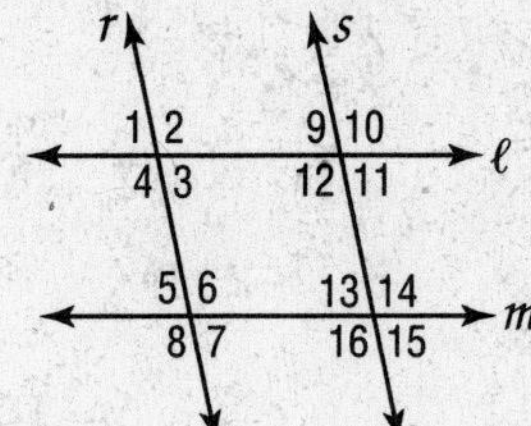

Statements	Reasons
1. $\angle 15 \cong \angle 5$	1. ________________
2. $\angle 13 \cong \angle 15$	2. ________________
3. $\angle 5 \cong \angle 13$	3. ________________
4. $r \parallel s$	4. ________________
5. ________________	5. Given
6. ________________	6. If corr $\angle$s are $\cong$, then lines $\parallel$.

7. Determine whether $\overleftrightarrow{PQ} \perp \overleftrightarrow{TQ}$. Explain why or why not.

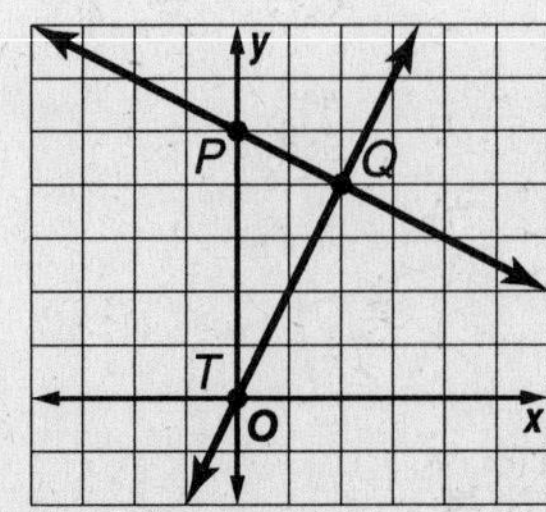

3-6 Study Guide and Intervention

Perpendiculars and Distance

Distance From a Point to a Line When a point is not on a line, the distance from the point to the line is the length of the segment that contains the point and is perpendicular to the line.

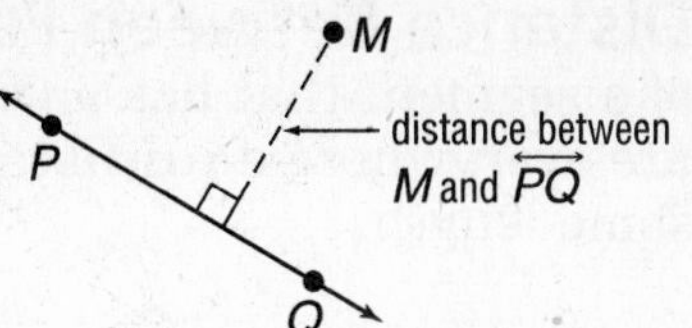

Example **Draw the segment that represents the distance from E to $\overleftrightarrow{AF}$.**

Extend $\overleftrightarrow{AF}$. Draw $\overline{EG} \perp \overleftrightarrow{AF}$.

$\overline{EG}$ represents the distance from E to $\overleftrightarrow{AF}$.

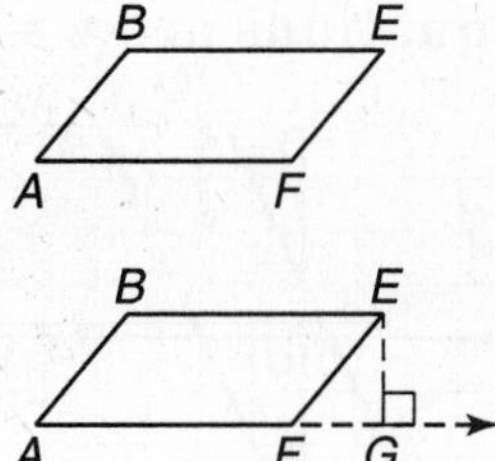

Exercises

Draw the segment that represents the distance indicated.

1. C to $\overleftrightarrow{AB}$

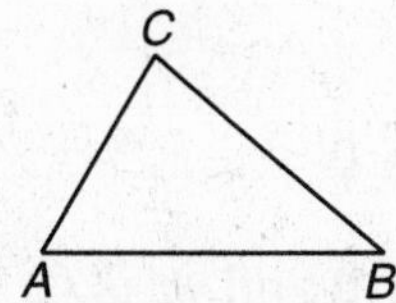

2. D to $\overleftrightarrow{AB}$

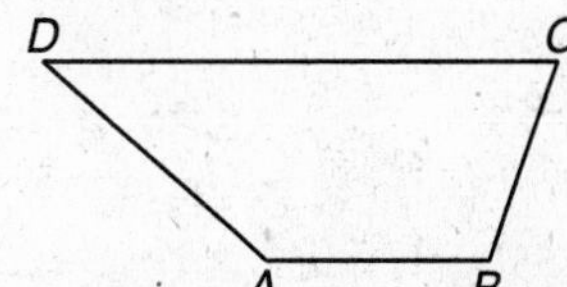

3. T to $\overleftrightarrow{RS}$

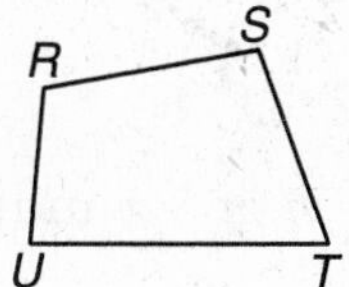

4. S to $\overleftrightarrow{PQ}$

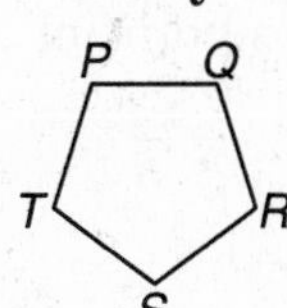

5. S to $\overleftrightarrow{QR}$

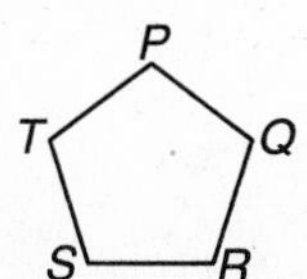

6. S to $\overleftrightarrow{RT}$

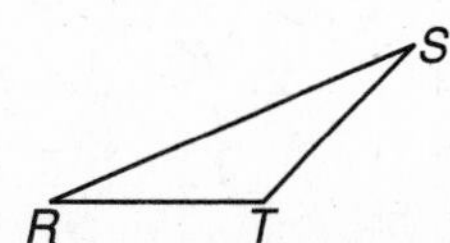

3-6 Study Guide and Intervention *(continued)*

Perpendiculars and Distance

Distance Between Parallel Lines The distance between parallel lines is the length of a segment that has an endpoint on each line and is perpendicular to them. Parallel lines are everywhere **equidistant**, which means that all such perpendicular segments have the same length.

Example **Find the distance between the parallel lines ℓ and m whose equations are $y = 2x + 1$ and $y = 2x - 4$, respectively.**

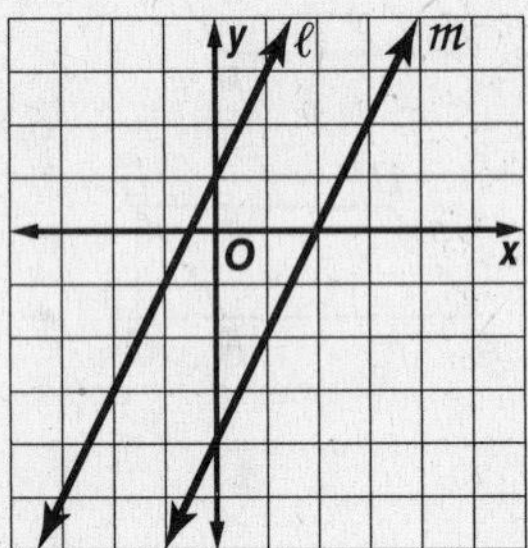

Draw a line p through (0, 1) that is perpendicular to ℓ and m.

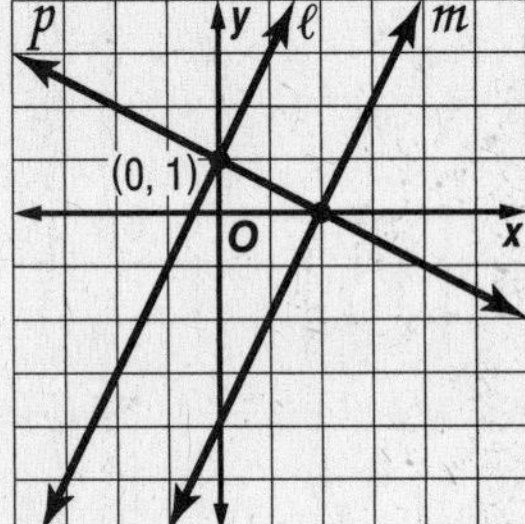

Line p has slope $-\frac{1}{2}$ and y-intercept 1. An equation of p is $y = -\frac{1}{2}x + 1$. The point of intersection for p and ℓ is (0, 1).

To find the point of intersection of p and m, solve a system of equations.

Line m: $y = 2x - 4$
Line p: $y = -\frac{1}{2}x + 1$

Use substitution.

$$2x - 4 = -\frac{1}{2}x + 1$$
$$4x - 8 = -x + 2$$
$$5x = 10$$
$$x = 2$$

Substitute 2 for x to find the y-coordinate.

$$y = -\frac{1}{2}x + 1$$
$$= -\frac{1}{2}(2) + 1 = -1 + 1 = 0$$

The point of intersection of p and m is (2, 0).

Use the Distance Formula to find the distance between (0, 1) and (2, 0).

$$d = \sqrt{(x_2 - x_1)^2 + (y_2 - y_1)^2}$$
$$= \sqrt{(2 - 0)^2 + (0 - 1)^2}$$
$$= \sqrt{5}$$

The distance between ℓ and m is $\sqrt{5}$ units.

Exercises

Find the distance between each pair of parallel lines.

1. $y = 8$
$y = -3$

2. $y = x + 3$
$y = x - 1$

3. $y = -2x$
$y = -2x - 5$

4-1 Study Guide and Intervention

Classifying Triangles

Classify Triangles by Angles One way to classify a triangle is by the measures of its angles.

• If *one* of the angles of a triangle is an obtuse angle, then the triangle is an **obtuse triangle**.
• If *one* of the angles of a triangle is a right angle, then the triangle is a **right triangle**.
• If *all three* of the angles of a triangle are acute angles, then the triangle is an **acute triangle**.
• If all three angles of an acute triangle are congruent, then the triangle is an **equiangular triangle**.

Example **Classify each triangle.**

a.

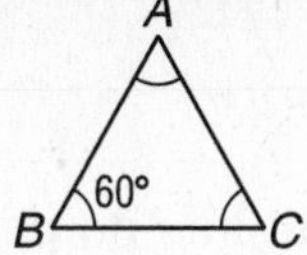

All three angles are congruent, so all three angles have measure 60°. The triangle is an equiangular triangle.

b.

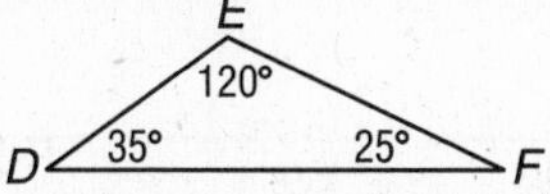

The triangle has one angle that is obtuse. It is an obtuse triangle.

c.

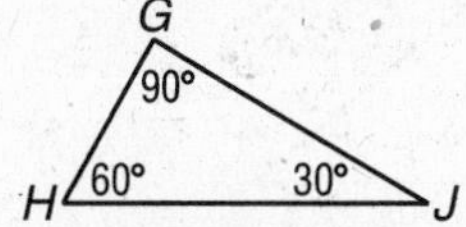

The triangle has one right angle. It is a right triangle.

Exercises

Classify each triangle as *acute*, *equiangular*, *obtuse*, or *right*.

1.

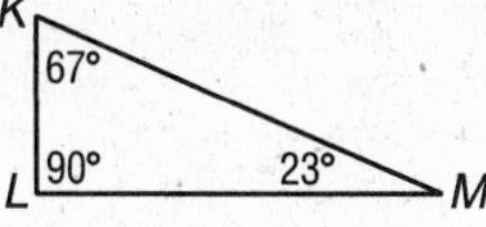

2.

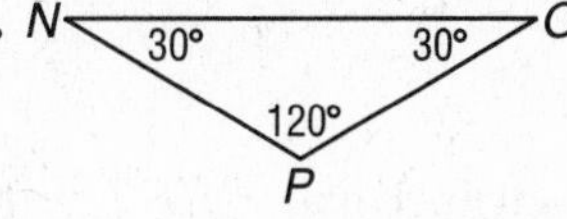

3.

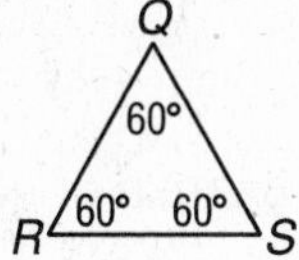

4.

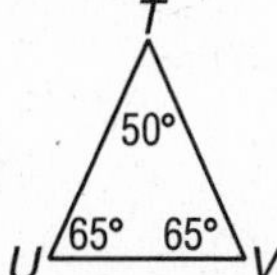

5.

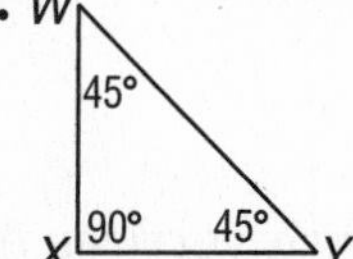

6.

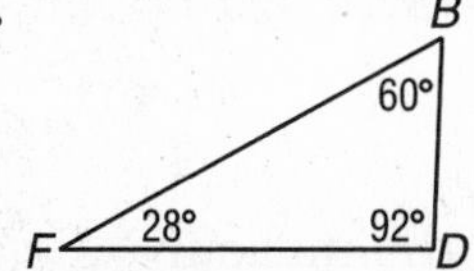

4-1 Study Guide and Intervention *(continued)*

Classifying Triangles

Classify Triangles by Sides You can classify a triangle by the measures of its sides. Equal numbers of hash marks indicate congruent sides.

• If *all three* sides of a triangle are congruent, then the triangle is an **equilateral triangle**.
• If *at least two* sides of a triangle are congruent, then the triangle is an **isosceles triangle**.
• If *no two* sides of a triangle are congruent, then the triangle is a **scalene triangle**.

Example **Classify each triangle.**

a.

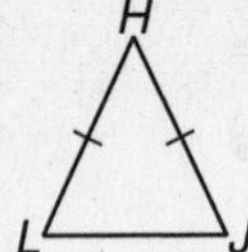

Two sides are congruent. The triangle is an isosceles triangle.

b.

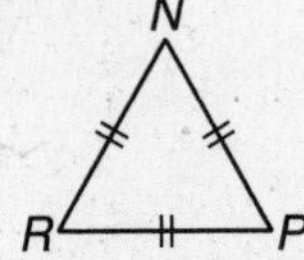

All three sides are congruent. The triangle is an equilateral triangle.

c.

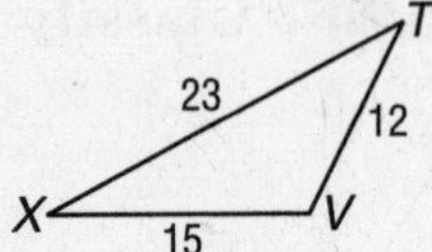

The triangle has no pair of congruent sides. It is a scalene triangle.

Exercises

Classify each triangle as *equilateral, isosceles,* or *scalene.*

1.

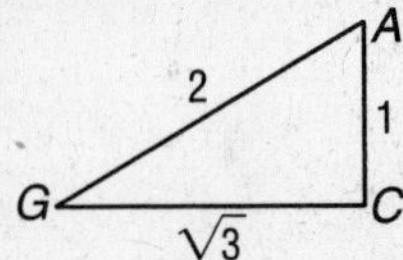

2.

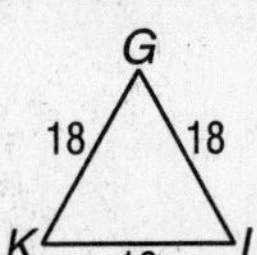

3.

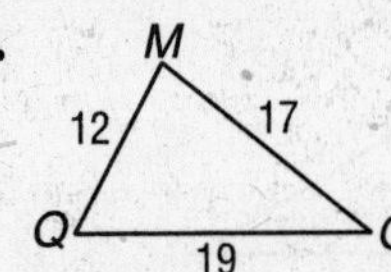

4.

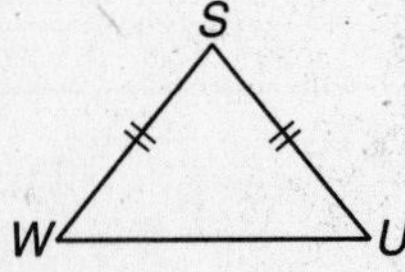

5.

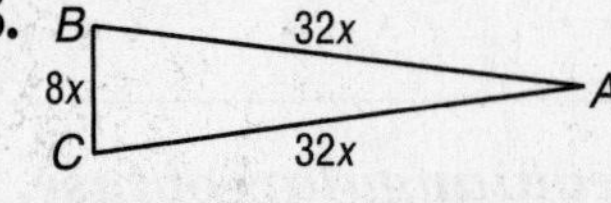

6.

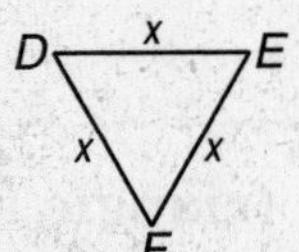

7. Find the measure of each side of equilateral $\triangle RST$ with $RS = 2x + 2$, $ST = 3x$, and $TR = 5x - 4$.

8. Find the measure of each side of isosceles $\triangle ABC$ with $AB = BC$ if $AB = 4y$, $BC = 3y + 2$, and $AC = 3y$.

9. Find the measure of each side of $\triangle ABC$ with vertices $A(-1, 5)$, $B(6, 1)$, and $C(2, -6)$. Classify the triangle.

NAME _______________ DATE ________ PERIOD ______

4-2 Study Guide and Intervention

Angles of Triangles

Angle Sum Theorem If the measures of two angles of a triangle are known, the measure of the third angle can always be found.

Angle Sum Theorem	The sum of the measures of the angles of a triangle is 180. In the figure at the right, $m\angle A + m\angle B + m\angle C = 180$. 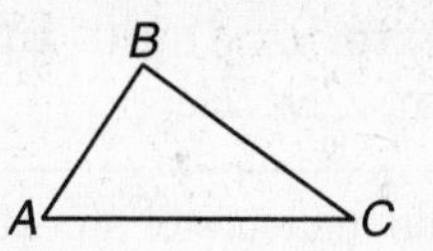

Example 1 **Find $m\angle T$.**

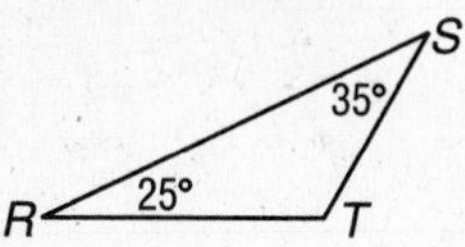

$m\angle R + m\angle S + m\angle T = 180$	Angle Sum Theorem
$25 + 35 + m\angle T = 180$	Substitution
$60 + m\angle T = 180$	Add.
$m\angle T = 120$	Subtract 60 from each side.

Example 2 **Find the missing angle measures.**

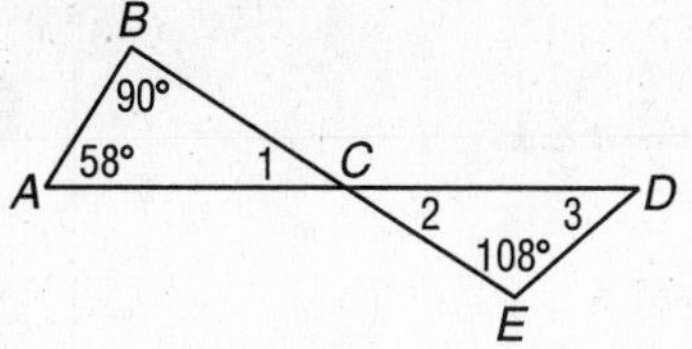

$m\angle 1 + m\angle A + m\angle B = 180$	Angle Sum Theorem
$m\angle 1 + 58 + 90 = 180$	Substitution
$m\angle 1 + 148 = 180$	Add.
$m\angle 1 = 32$	Subtract 148 from each side.
$m\angle 2 = 32$	Vertical angles are congruent.
$m\angle 3 + m\angle 2 + m\angle E = 180$	Angle Sum Theorem
$m\angle 3 + 32 + 108 = 180$	Substitution
$m\angle 3 + 140 = 180$	Add.
$m\angle 3 = 40$	Subtract 140 from each side.

Exercises

Find the measure of each numbered angle.

1.

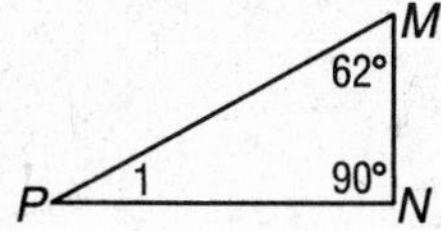

2.

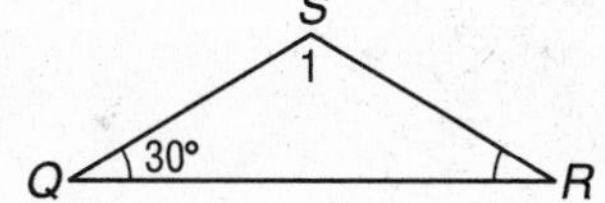

3.

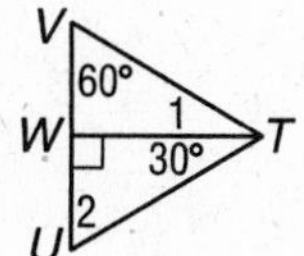

4.

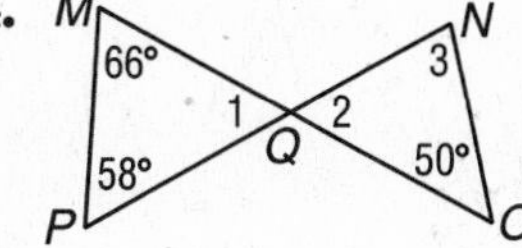

5.

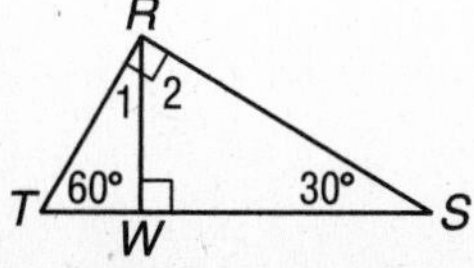

6.

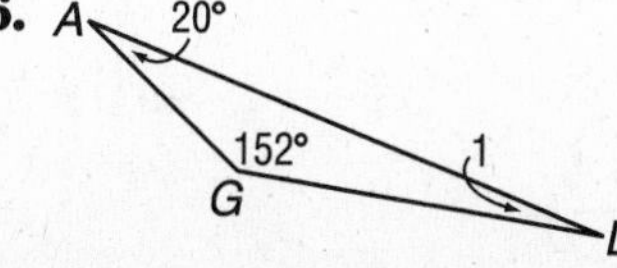

NAME ____________________ DATE __________ PERIOD _____

4-2 Study Guide and Intervention *(continued)*

Angles of Triangles

Exterior Angle Theorem At each vertex of a triangle, the angle formed by one side and an extension of the other side is called an **exterior angle** of the triangle. For each exterior angle of a triangle, the **remote interior angles** are the interior angles that are not adjacent to that exterior angle. In the diagram below, $\angle B$ and $\angle A$ are the remote interior angles for exterior $\angle DCB$.

Exterior Angle Theorem	The measure of an exterior angle of a triangle is equal to the sum of the measures of the two remote interior angles. $m\angle 1 = m\angle A + m\angle B$ 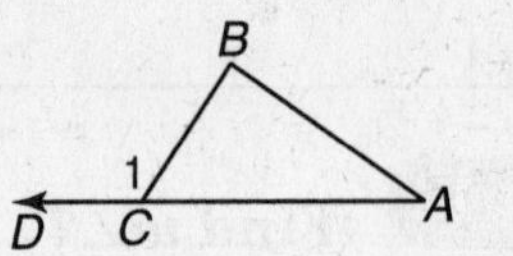

Example 1 **Find $m\angle 1$.**

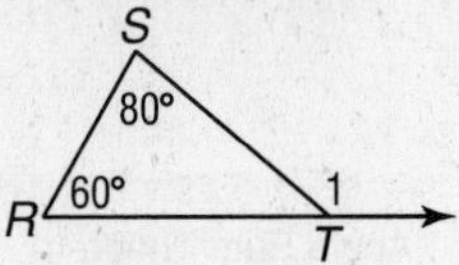

$m\angle 1 = m\angle R + m\angle S$ Exterior Angle Theorem

$= 60 + 80$ Substitution

$= 140$ Add.

Example 2 **Find x.**

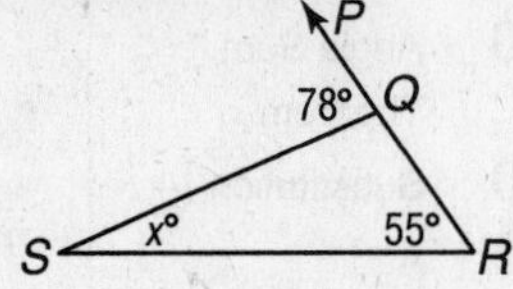

$m\angle PQS = m\angle R + m\angle S$ Exterior Angle Theorem

$78 = 55 + x$ Substitution

$23 = x$ Subtract 55 from each side.

Exercises

Find the measure of each numbered angle.

1.

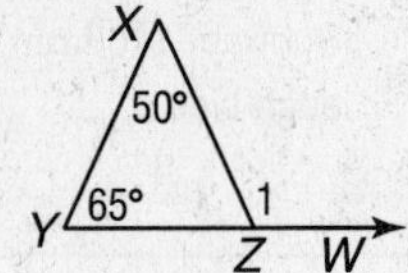

2.

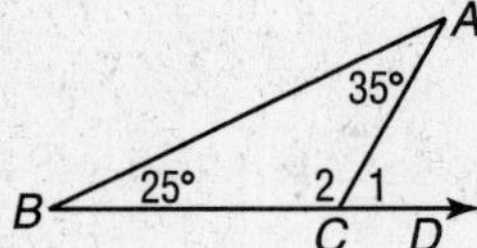

3.

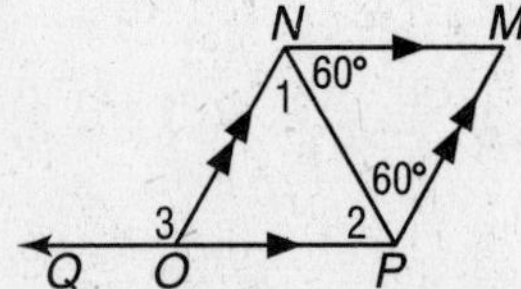

4.

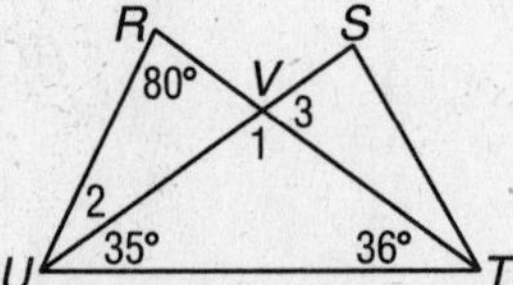

Find x.

5.

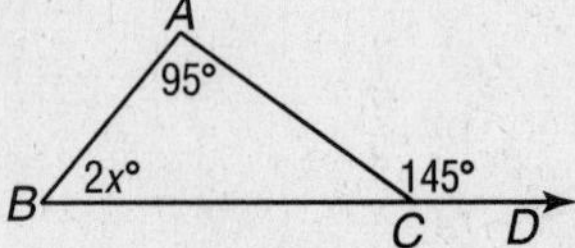

6.

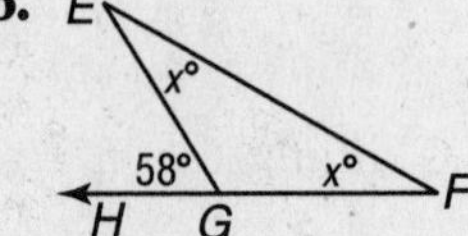

4-3 Study Guide and Intervention

Congruent Triangles

Corresponding Parts of Congruent Triangles

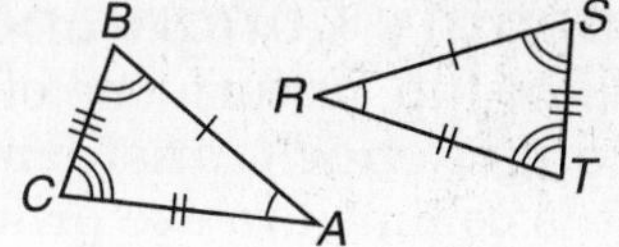

Triangles that have the same size and same shape are **congruent triangles**. Two triangles are congruent if and only if all three pairs of corresponding angles are congruent and all three pairs of corresponding sides are congruent. In the figure, $\triangle ABC \cong \triangle RST$.

Example **If $\triangle XYZ \cong \triangle RST$, name the pairs of congruent angles and congruent sides.**

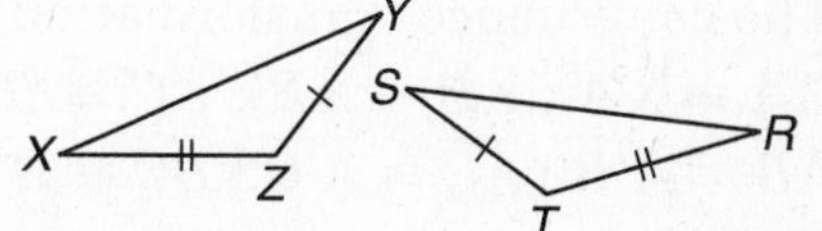

$\angle X \cong \angle R$, $\angle Y \cong \angle S$, $\angle Z \cong \angle T$

$\overline{XY} \cong \overline{RS}$, $\overline{XZ} \cong \overline{RT}$, $\overline{YZ} \cong \overline{ST}$

Exercises

Identify the congruent triangles in each figure.

1. 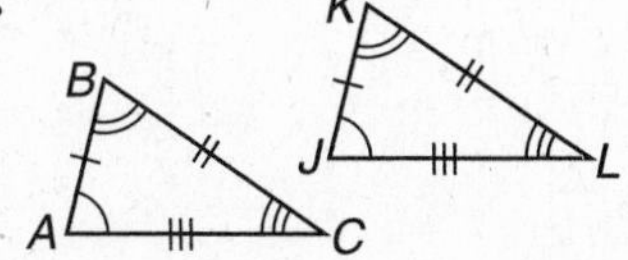

2. 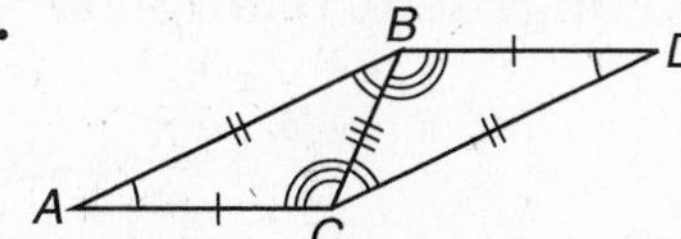

3. 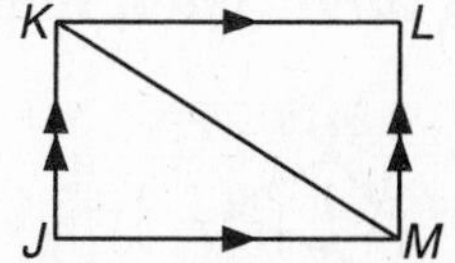

Name the corresponding congruent angles and sides for the congruent triangles.

4. 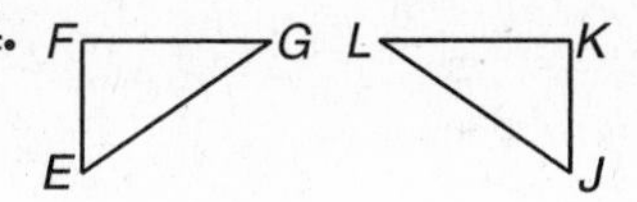

5. 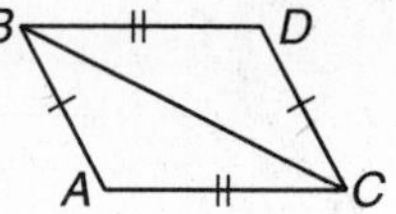

6. 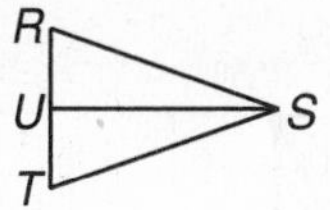

4-3 Study Guide and Intervention *(continued)*

Congruent Triangles

Identify Congruence Transformations If two triangles are congruent, you can slide, flip, or turn one of the triangles and they will still be congruent. These are called **congruence transformations** because they do not change the size or shape of the figure. It is common to use prime symbols to distinguish between an original $\triangle ABC$ and a transformed $\triangle A'B'C'$.

Example **Name the congruence transformation that produces $\triangle A'B'C'$ from $\triangle ABC$.**

The congruence transformation is a slide.

$\angle A \cong \angle A'$; $\angle B \cong \angle B'$; $\angle C \cong \angle C'$;
$\overline{AB} \cong \overline{A'B'}$; $\overline{AC} \cong \overline{A'C'}$; $\overline{BC} \cong \overline{B'C'}$

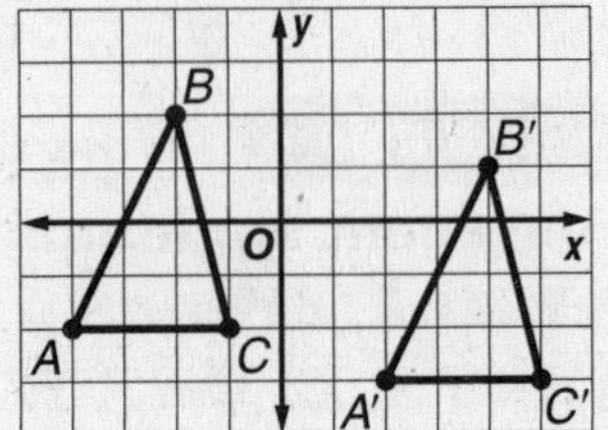

Exercises

Describe the congruence transformation between the two triangles as a *slide*, a *flip*, or a *turn*. Then name the congruent triangles.

1.
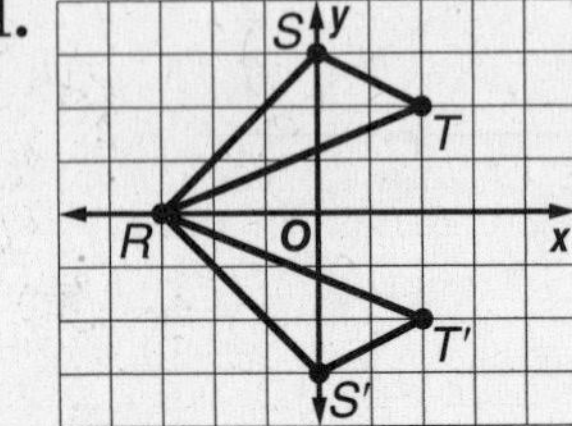

2.
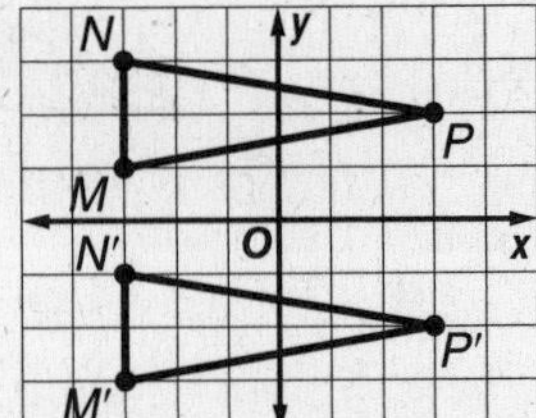

3.
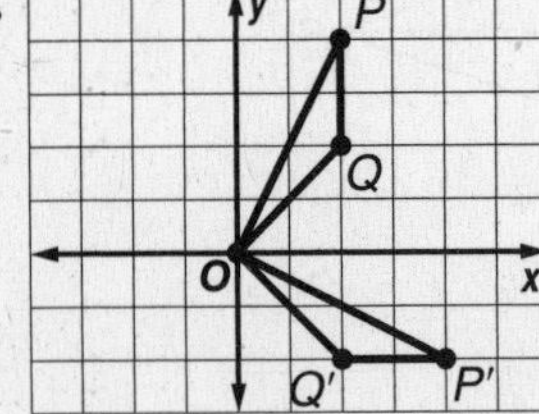

4.
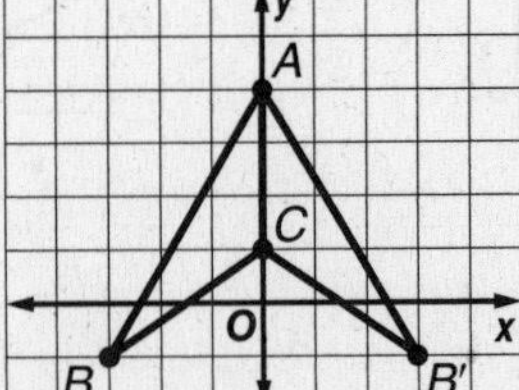

5.
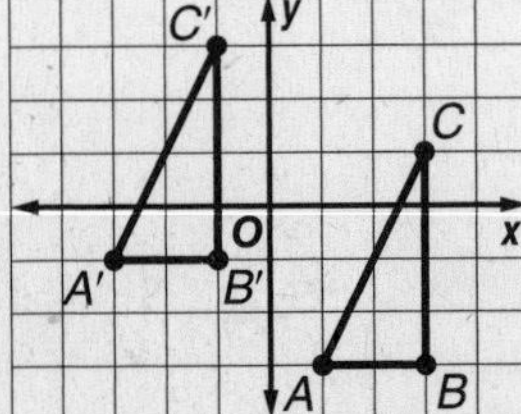

6.
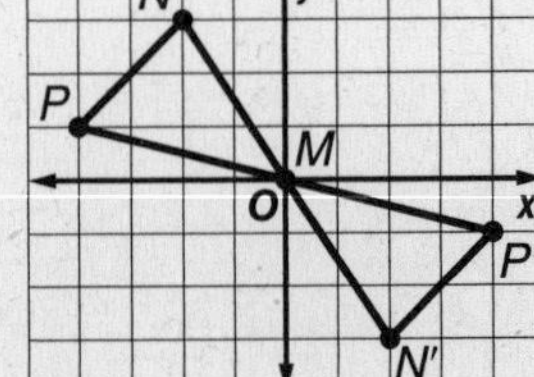

4-4 Study Guide and Intervention

Proving Congruence—SSS, SAS

SSS Postulate You know that two triangles are congruent if corresponding sides are congruent and corresponding angles are congruent. The Side-Side-Side (SSS) Postulate lets you show that two triangles are congruent if you know only that the sides of one triangle are congruent to the sides of the second triangle.

SSS Postulate	If the sides of one triangle are congruent to the sides of a second triangle, then the triangles are congruent.

Example **Write a two-column proof.**

Given: $\overline{AB} \cong \overline{DB}$ and C is the midpoint of $\overline{AD}$.
Prove: $\triangle ABC \cong \triangle DBC$

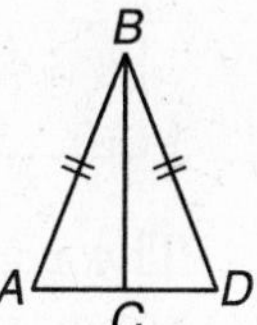

Statements	Reasons
1. $\overline{AB} \cong \overline{DB}$	1. Given
2. C is the midpoint of $\overline{AD}$.	2. Given
3. $\overline{AC} \cong \overline{DC}$	3. Definition of midpoint
4. $\overline{BC} \cong \overline{BC}$	4. Reflexive Property of ≅
5. $\triangle ABC \cong \triangle DBC$	5. SSS Postulate

Exercises

Write a two-column proof.

1.

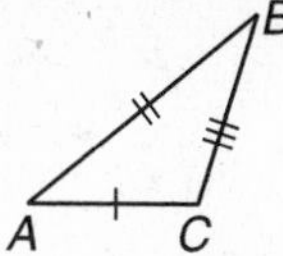

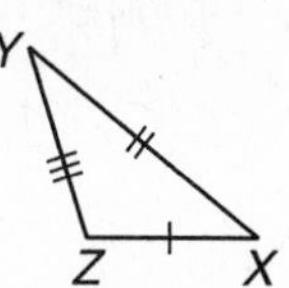

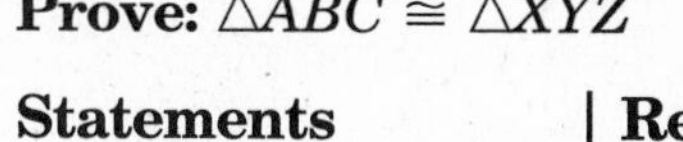

Given: $\overline{AB} \cong \overline{XY}$, $\overline{AC} \cong \overline{XZ}$, $\overline{BC} \cong \overline{YZ}$
Prove: $\triangle ABC \cong \triangle XYZ$

Statements	Reasons

2.

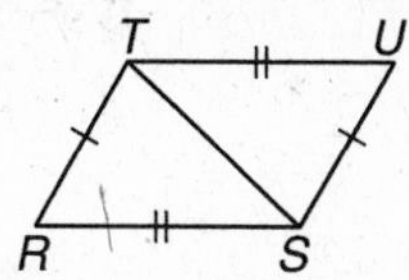

Given: $\overline{RS} \cong \overline{UT}$, $\overline{RT} \cong \overline{US}$
Prove: $\triangle RST \cong \triangle UTS$

Statements	Reasons

Lesson 4-4

4-4 Study Guide and Intervention *(continued)*

Proving Congruence—SSS, SAS

SAS Postulate Another way to show that two triangles are congruent is to use the Side-Angle-Side (SAS) Postulate.

SAS Postulate	If two sides and the included angle of one triangle are congruent to two sides and the included angle of another triangle, then the triangles are congruent.

Example **For each diagram, determine which pairs of triangles can be proved congruent by the SAS Postulate.**

a.

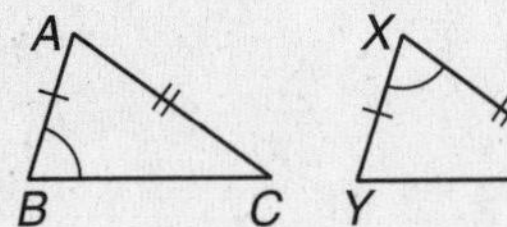

In $\triangle ABC$, the angle is not "included" by the sides $\overline{AB}$ and $\overline{AC}$. So the triangles cannot be proved congruent by the SAS Postulate.

b.

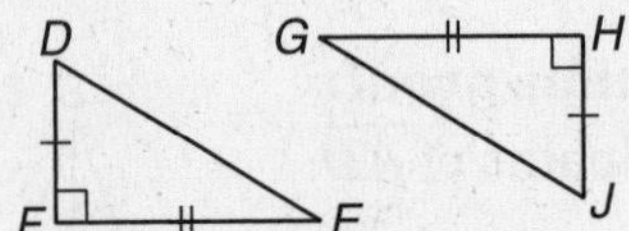

The right angles are congruent and they are the included angles for the congruent sides. $\triangle DEF \cong \triangle JGH$ by the SAS Postulate.

c.

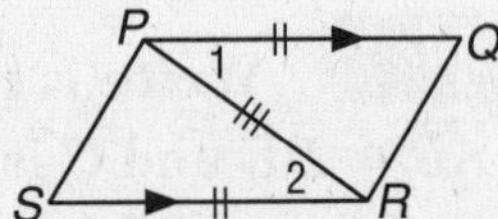

The included angles, $\angle 1$ and $\angle 2$, are congruent because they are alternate interior angles for two parallel lines. $\triangle PSR \cong \triangle RQP$ by the SAS Postulate.

Exercises

For each figure, determine which pairs of triangles can be proved congruent by the SAS Postulate.

1.

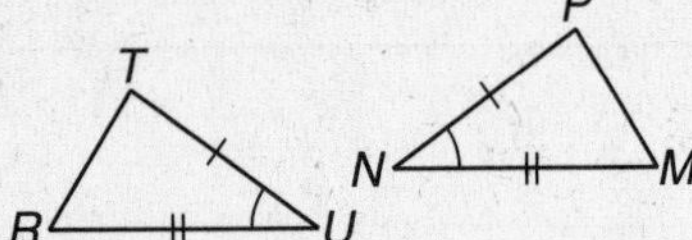

2.

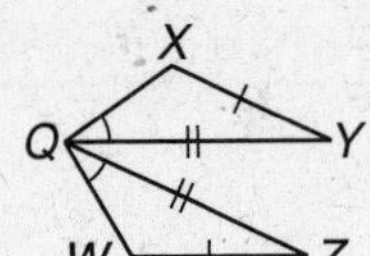

3.

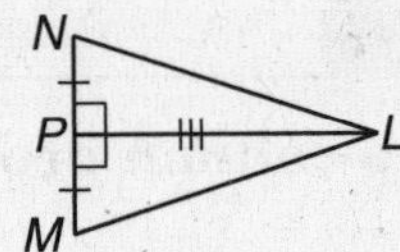

4.

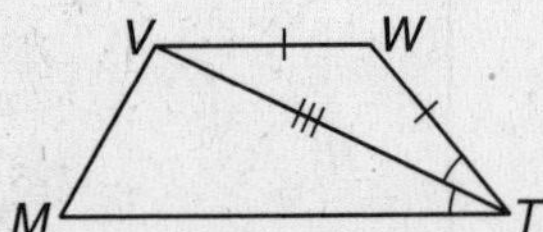

5.

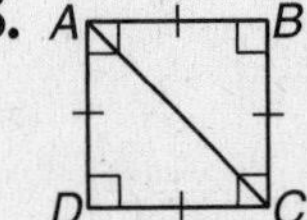

6.

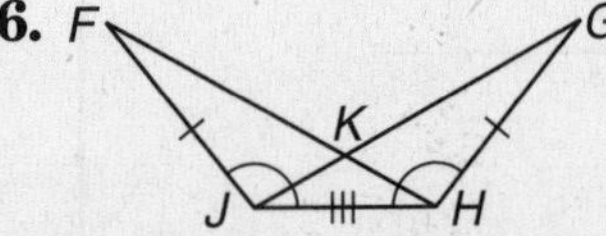

4-5 Study Guide and Intervention

Proving Congruence—ASA, AAS

ASA Postulate The Angle-Side-Angle (ASA) Postulate lets you show that two triangles are congruent.

ASA Postulate	If two angles and the included side of one triangle are congruent to two angles and the included side of another triangle, then the triangles are congruent.

Example **Find the missing congruent parts so that the triangles can be proved congruent by the ASA Postulate. Then write the triangle congruence.**

a.

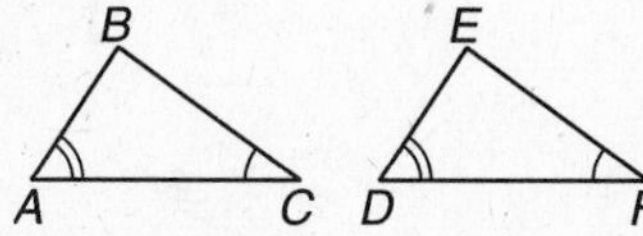

Two pairs of corresponding angles are congruent, $\angle A \cong \angle D$ and $\angle C \cong \angle F$. If the included sides $\overline{AC}$ and $\overline{DF}$ are congruent, then $\triangle ABC \cong \triangle DEF$ by the ASA Postulate.

b.

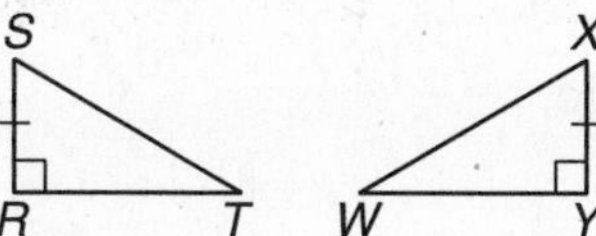

$\angle R \cong \angle Y$ and $\overline{SR} \cong \overline{XY}$. If $\angle S \cong \angle X$, then $\triangle RST \cong \triangle YXW$ by the ASA Postulate.

Exercises

What corresponding parts must be congruent in order to prove that the triangles are congruent by the ASA Postulate? Write the triangle congruence statement.

1.

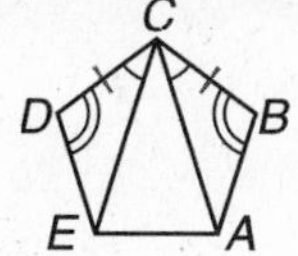

2.

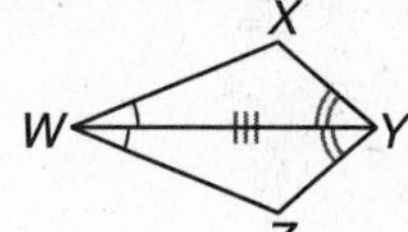

3.

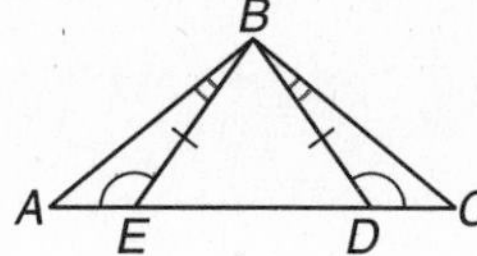

4.

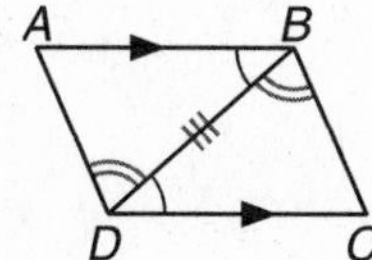

5.

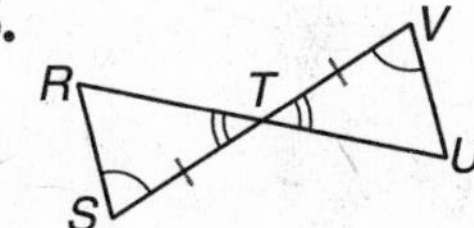

6.

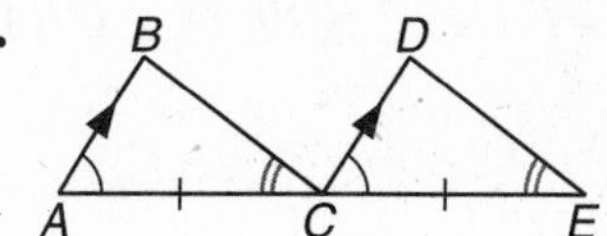

Lesson 4-5

NAME ______________________________ DATE ____________ PERIOD _____

4-5 Study Guide and Intervention *(continued)*

Proving Congruence—ASA, AAS

AAS Theorem Another way to show that two triangles are congruent is the Angle-Angle-Side (AAS) Theorem.

AAS Theorem	If two angles and a nonincluded side of one triangle are congruent to the corresponding two angles and side of a second triangle, then the two triangles are congruent.

You now have five ways to show that two triangles are congruent.

- definition of triangle congruence
- SSS Postulate
- SAS Postulate
- ASA Postulate
- AAS Theorem

Example **In the diagram, $\angle BCA \cong \angle DCA$. Which sides are congruent? Which additional pair of corresponding parts needs to be congruent for the triangles to be congruent by the AAS Postulate?**

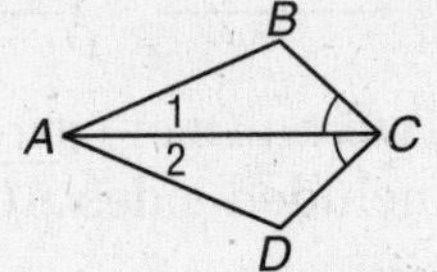

$\overline{AC} \cong \overline{AC}$ by the Reflexive Property of congruence. The congruent angles cannot be $\angle 1$ and $\angle 2$, because $\overline{AC}$ would be the included side. If $\angle B \cong \angle D$, then $\triangle ABC \cong \triangle ADC$ by the AAS Theorem.

Exercises

In Exercises 1 and 2, draw and label $\triangle ABC$ and $\triangle DEF$. Indicate which additional pair of corresponding parts needs to be congruent for the triangles to be congruent by the AAS Theorem.

1. $\angle A \cong \angle D$; $\angle B \cong \angle E$

2. $BC \cong EF$; $\angle A \cong \angle D$

3. Write a flow proof.

Given: $\angle S \cong \angle U$; $\overline{TR}$ bisects $\angle STU$.

Prove: $\angle SRT \cong \angle URT$

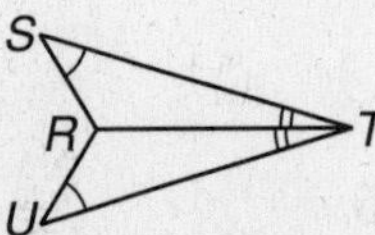

NAME ______________________ DATE __________ PERIOD _____

4-6 Study Guide and Intervention

Isosceles Triangles

Lesson 4-6

Properties of Isosceles Triangles An **isosceles triangle** has two congruent sides. The angle formed by these sides is called the **vertex angle**. The other two angles are called **base angles**. You can prove a theorem and its converse about isosceles triangles.

- If two sides of a triangle are congruent, then the angles opposite those sides are congruent. **(Isosceles Triangle Theorem)**
- If two angles of a triangle are congruent, then the sides opposite those angles are congruent.

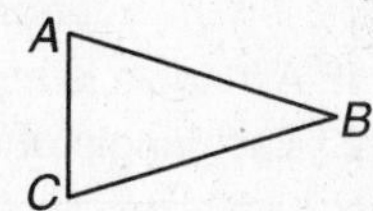

If $\overline{AB} \cong \overline{CB}$, then $\angle A \cong \angle C$.

If $\angle A \cong \angle C$, then $\overline{AB} \cong \overline{CB}$.

Example 1 **Find x, given $\overline{BC} \cong \overline{BA}$.**

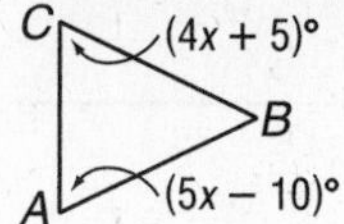

$BC = BA$, so

$m\angle A = m\angle C$.	Isos. Triangle Theorem
$5x - 10 = 4x + 5$	Substitution
$x - 10 = 5$	Subtract 4*x* from each side.
$x = 15$	Add 10 to each side.

Example 2 **Find x.**

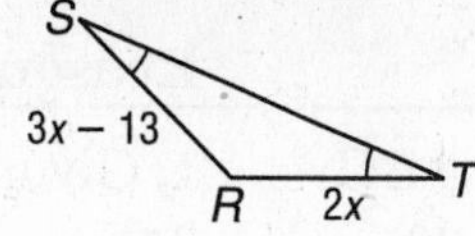

$m\angle S = m\angle T$, so

$SR = TR$.	Converse of Isos. △ Thm.
$3x - 13 = 2x$	Substitution
$3x = 2x + 13$	Add 13 to each side.
$x = 13$	Subtract 2*x* from each side.

Exercises

Find x.

1.

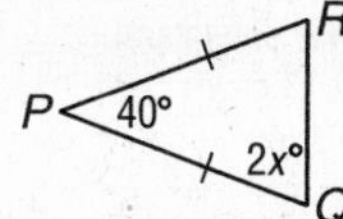

2.

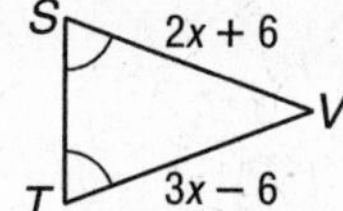

3.

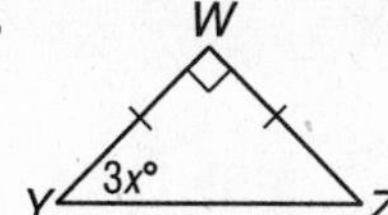

4.

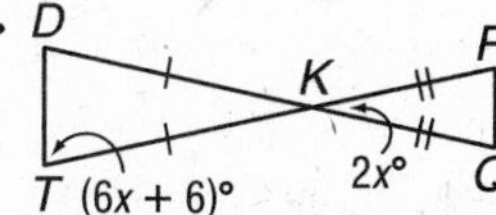

5.

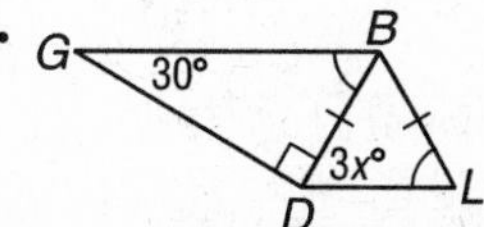

6.

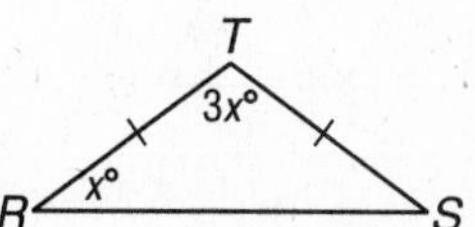

7. Write a two-column proof.

Given: $\angle 1 \cong \angle 2$

Prove: $\overline{AB} \cong \overline{CB}$

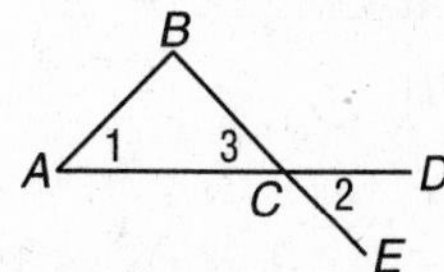

Statements	**Reasons**

4-6 Study Guide and Intervention *(continued)*

Isosceles Triangles

Properties of Equilateral Triangles An **equilateral triangle** has three congruent sides. The Isosceles Triangle Theorem can be used to prove two properties of equilateral triangles.

1. A triangle is equilateral if and only if it is equiangular.
2. Each angle of an equilateral triangle measures 60°.

Example **Prove that if a line is parallel to one side of an equilateral triangle, then it forms another equilateral triangle.**

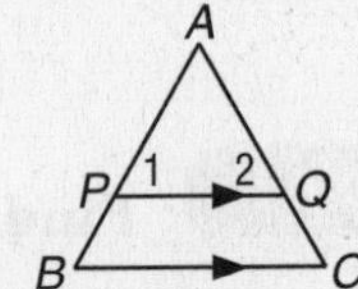

Proof:

Statements	**Reasons**
1. $\triangle ABC$ is equilateral; $\overline{PQ} \parallel \overline{BC}$.	1. Given
2. $m\angle A = m\angle B = m\angle C = 60$	2. Each $\angle$ of an equilateral $\triangle$ measures 60°.
3. $\angle 1 \cong \angle B$, $\angle 2 \cong \angle C$	3. If $\parallel$ lines, then corres. $\angle$s are $\cong$.
4. $m\angle 1 = 60$, $m\angle 2 = 60$	4. Substitution
5. $\triangle APQ$ is equilateral.	5. If a $\triangle$ is equiangular, then it is equilateral.

Exercises

Find x.

1.

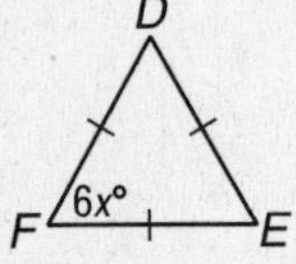

2.

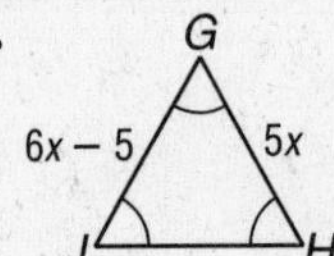

3.

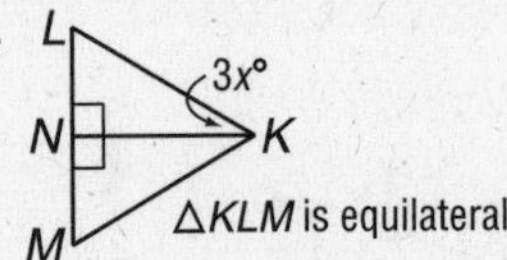

4.

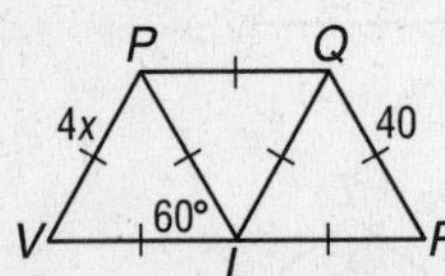

5.

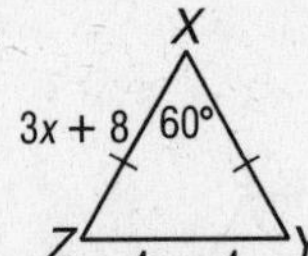

6.

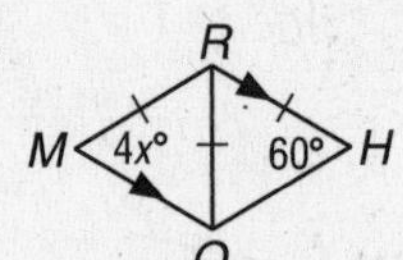

7. Write a two-column proof.

Given: $\triangle ABC$ is equilateral; $\angle 1 \cong \angle 2$.

Prove: $\angle ADB \cong \angle CDB$

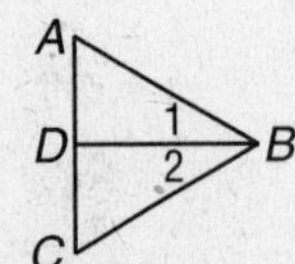

Proof:

Statements	**Reasons**

4-7 Study Guide and Intervention

Triangles and Coordinate Proof

Position and Label Triangles A coordinate proof uses points, distances, and slopes to prove geometric properties. The first step in writing a coordinate proof is to place a figure on the coordinate plane and label the vertices. Use the following guidelines.

1. Use the origin as a vertex or center of the figure.
2. Place at least one side of the polygon on an axis.
3. Keep the figure in the first quadrant if possible.
4. Use coordinates that make the computations as simple as possible.

Example **Position an isosceles triangle on the coordinate plane so that its sides are a units long and one side is on the positive x-axis.**

Start with $R(0, 0)$. If RT is a, then another vertex is $T(a, 0)$.

For vertex S, the x-coordinate is $\frac{a}{2}$. Use b for the y-coordinate, so the vertex is $S\left(\frac{a}{2}, b\right)$.

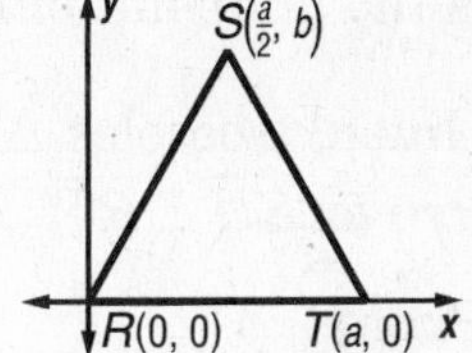

Exercises

Find the missing coordinates of each triangle.

1.
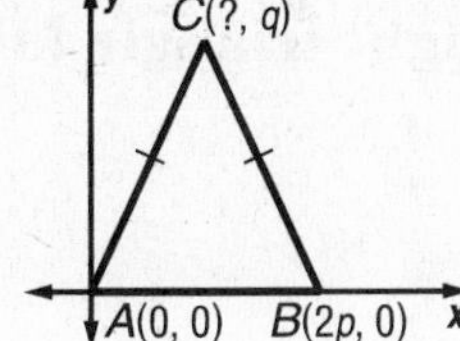

2.
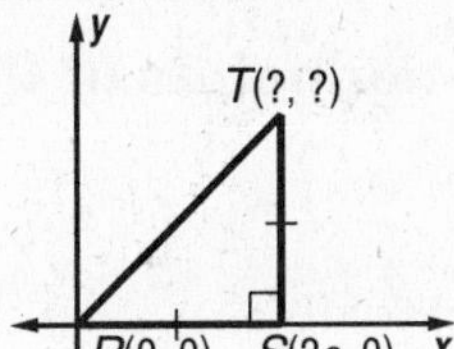

3.
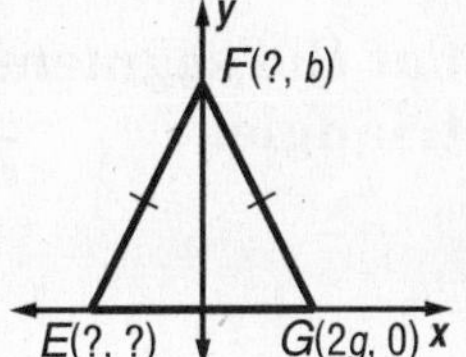

Position and label each triangle on the coordinate plane.

4. isosceles triangle $\triangle RST$ with base $\overline{RS}$ and $4a$ units long

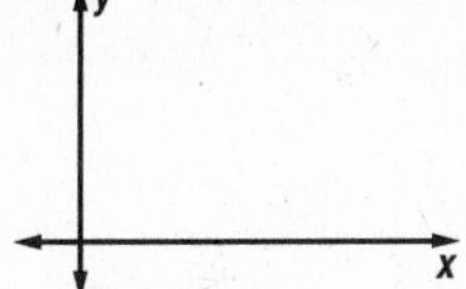

5. isosceles right $\triangle DEF$ with legs e units long

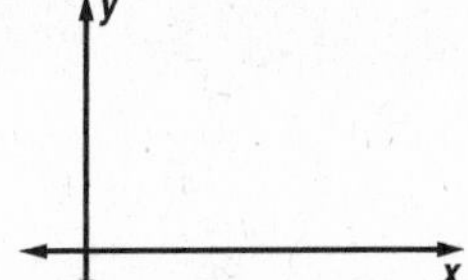

6. equilateral triangle $\triangle EQI$ with vertex $Q(0, \sqrt{3}b)$ sides $2b$ units long

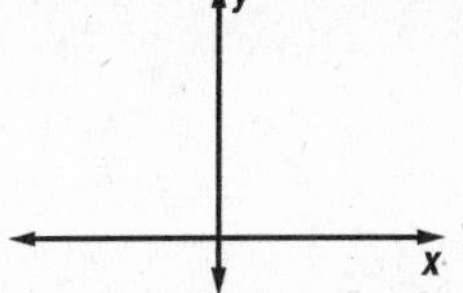

NAME ______________________ DATE ____________ PERIOD _____

4-7 Study Guide and Intervention *(continued)*

Triangles and Coordinate Proof

Write Coordinate Proofs Coordinate proofs can be used to prove theorems and to verify properties. Many coordinate proofs use the Distance Formula, Slope Formula, or Midpoint Theorem.

Example **Prove that a segment from the vertex angle of an isosceles triangle to the midpoint of the base is perpendicular to the base.**

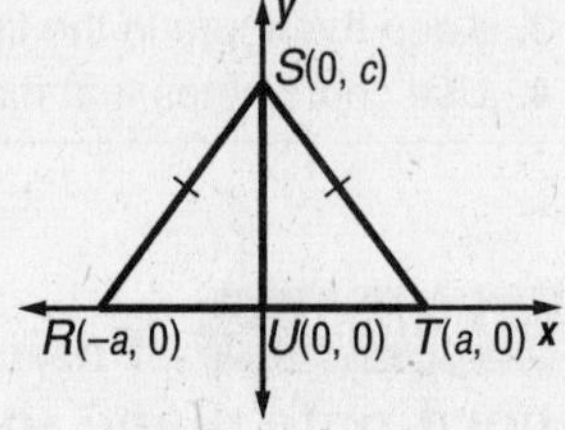

First, position and label an isosceles triangle on the coordinate plane. One way is to use $T(a, 0)$, $R(-a, 0)$, and $S(0, c)$. Then $U(0, 0)$ is the midpoint of $\overline{RT}$.

Given: Isosceles $\triangle RST$; U is the midpoint of base $\overline{RT}$.
Prove: $\overline{SU} \perp \overline{RT}$

Proof:
U is the midpoint of $\overline{RT}$ so the coordinates of U are $\left(\frac{-a + a}{2}, \frac{0 + 0}{2}\right) = (0, 0)$. Thus $\overline{SU}$ lies on the y-axis, and $\triangle RST$ was placed so $\overline{RT}$ lies on the x-axis. The axes are perpendicular, so $\overline{SU} \perp \overline{RT}$.

Exercises

Prove that the segments joining the midpoints of the sides of a right triangle form a right triangle.

© Glencoe/McGraw-Hill

NAME ______________________ DATE __________ PERIOD ____

5-1 Study Guide and Intervention

Bisectors, Medians, and Altitudes

Perpendicular Bisectors and Angle Bisectors A **perpendicular bisector** of a side of a triangle is a line, segment, or ray that is perpendicular to the side and passes through its midpoint. Another special segment, ray, or line is an **angle bisector**, which divides an angle into two congruent angles.

Two properties of perpendicular bisectors are:

(1) a point is on the perpendicular bisector of a segment if and only if it is equidistant from the endpoints of the segment, and

(2) the three perpendicular bisectors of the sides of a triangle meet at a point, called the **circumcenter** of the triangle, that is equidistant from the three vertices of the triangle.

Two properties of angle bisectors are:

(1) a point is on the angle bisector of an angle if and only if it is equidistant from the sides of the angle, and

(2) the three angle bisectors of a triangle meet at a point, called the **incenter** of the triangle, that is equidistant from the three sides of the triangle.

Lesson 5-1

Example 1 $\overrightarrow{BD}$ **is the perpendicular bisector of** $\overline{AC}$**. Find** x**.**

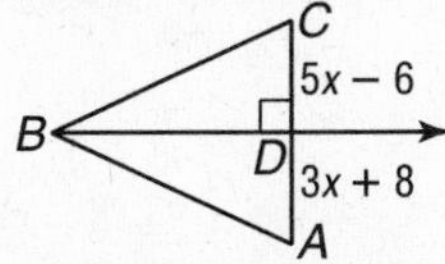

$\overrightarrow{BD}$ is the perpendicular bisector of $\overline{AC}$, so $AD = DC$.

$3x + 8 = 5x - 6$

$14 = 2x$

$7 = x$

Example 2 $\overrightarrow{MR}$ **is the angle bisector of** $\angle NMP$**. Find** x **if** $m\angle 1 = 5x + 8$ **and** $m\angle 2 = 8x - 16$**.**

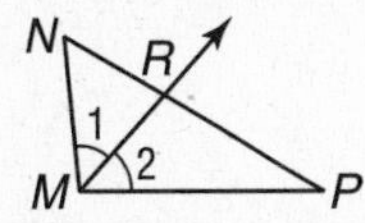

$\overrightarrow{MR}$ is the angle bisector of $\angle NMP$, so $m\angle 1 = m\angle 2$.

$5x + 8 = 8x - 16$

$24 = 3x$

$8 = x$

Exercises

Find the value of each variable.

1.

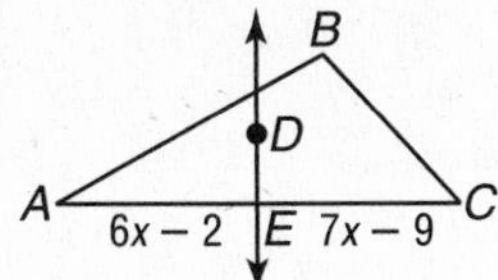

$\overleftrightarrow{DE}$ is the perpendicular bisector of $\overline{AC}$.

2.

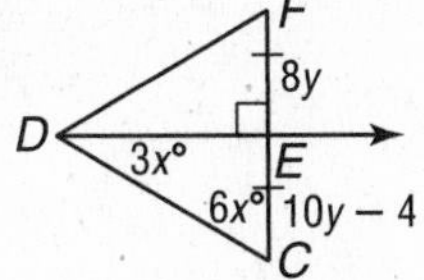

$\triangle CDF$ is equilateral.

3.

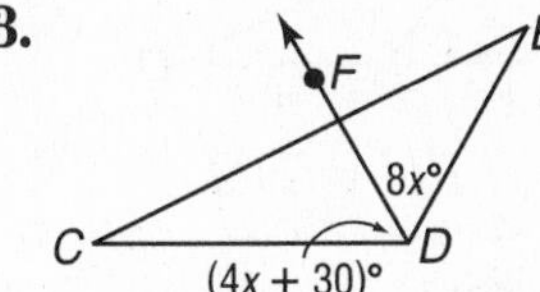

$\overrightarrow{DF}$ bisects $\angle CDE$.

4. For what kinds of triangle(s) can the perpendicular bisector of a side also be an angle bisector of the angle opposite the side?

5. For what kind of triangle do the perpendicular bisectors intersect in a point outside the triangle?

NAME ______________________________ DATE __________ PERIOD ____

5-1 Study Guide and Intervention *(continued)*

Bisectors, Medians, and Altitudes

Medians and Altitudes A **median** is a line segment that connects the vertex of a triangle to the midpoint of the opposite side. The three medians of a triangle intersect at the **centroid** of the triangle.

Centroid Theorem	The centroid of a triangle is located two thirds of the distance from a vertex to the midpoint of the side opposite the vertex on a median.

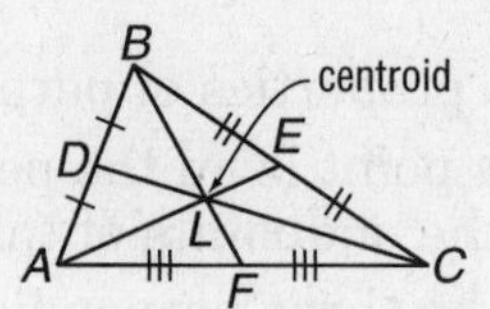

$AL = \frac{2}{3}AE,\ BL = \frac{2}{3}BF,\ CL = \frac{2}{3}CD$

Example **Points R, S, and T are the midpoints of $\overline{AB}$, $\overline{BC}$ and $\overline{AC}$, respectively. Find x, y, and z.**

$$CU = \frac{2}{3}CR \qquad BU = \frac{2}{3}BT \qquad AU = \frac{2}{3}AS$$

$$6x = \frac{2}{3}(6x + 15) \qquad 24 = \frac{2}{3}(24 + 3y - 3) \qquad 6z + 4 = \frac{2}{3}(6z + 4 + 11)$$

$$9x = 6x + 15 \qquad 36 = 24 + 3y - 3 \qquad \frac{3}{2}(6z + 4) = 6z + 4 + 11$$

$$3x = 15 \qquad 36 = 21 + 3y \qquad 9z + 6 = 6z + 15$$

$$x = 5 \qquad 15 = 3y \qquad 3z = 9$$

$$5 = y \qquad z = 3$$

Exercises

Find the value of each variable.

1.

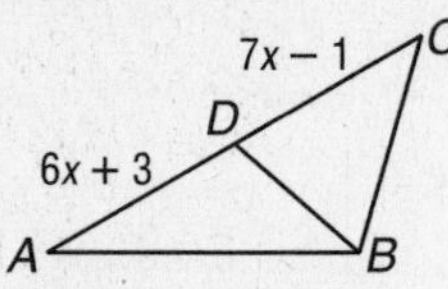

$\overline{BD}$ is a median.

2.

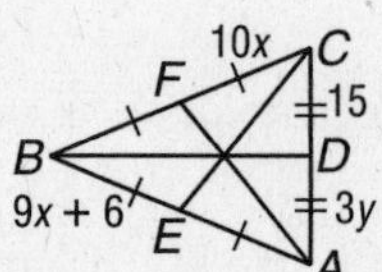

$AB = CB$; D, E, and F are midpoints.

3.

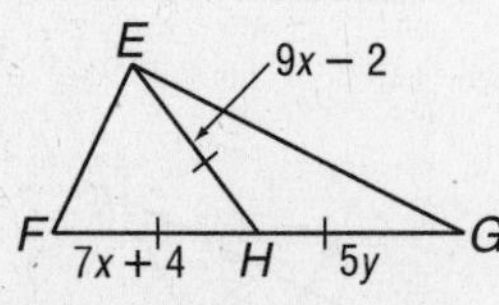

$EH = FH = HG$

4.

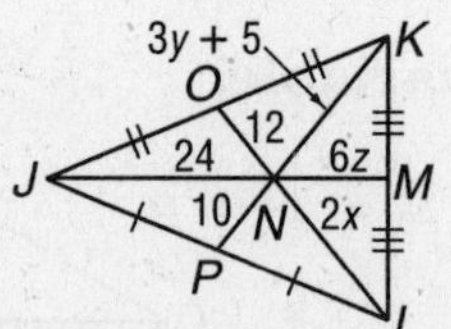

5.

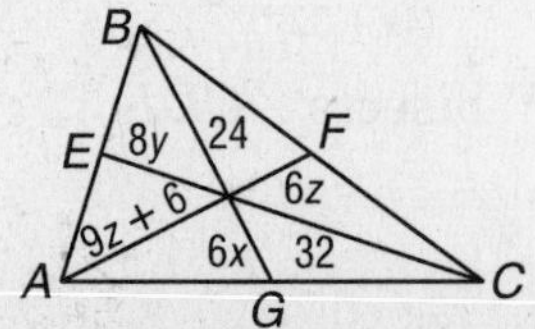

D is the centroid of $\triangle ABC$.

6.

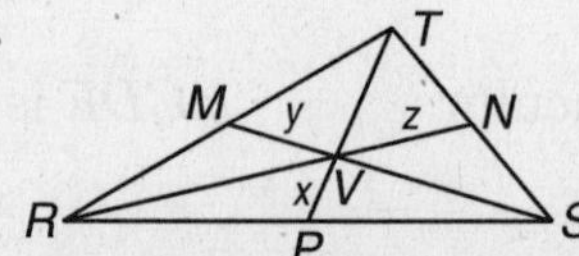

V is the centroid of $\triangle RST$;
$TP = 18$; $MS = 15$; $RN = 24$

7. For what kind of triangle are the medians and angle bisectors the same segments?

8. For what kind of triangle is the centroid outside the triangle?

© Glencoe/McGraw-Hill

NAME ______________________________ DATE ____________ PERIOD _____

5-2 Study Guide and Intervention

Inequalities and Triangles

Angle Inequalities Properties of inequalities, including the Transitive, Addition, Subtraction, Multiplication, and Division Properties of Inequality, can be used with measures of angles and segments. There is also a Comparison Property of Inequality.

For any real numbers a and b, either $a < b$, $a = b$, or $a > b$.

The Exterior Angle Theorem can be used to prove this inequality involving an exterior angle.

Exterior Angle Inequality Theorem	If an angle is an exterior angle of a triangle, then its measure is greater than the measure of either of its corresponding remote interior angles.	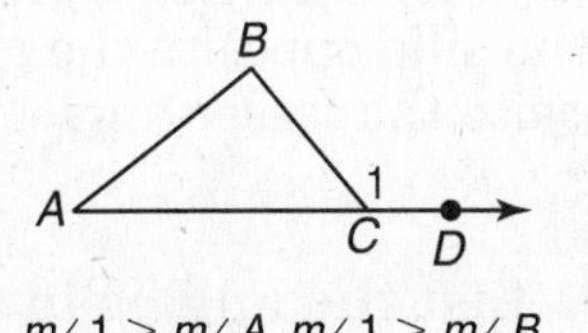 $m\angle 1 > m\angle A$, $m\angle 1 > m\angle B$

Example **List all angles of $\triangle EFG$ whose measures are less than $m\angle 1$.**

The measure of an exterior angle is greater than the measure of either remote interior angle. So $m\angle 3 < m\angle 1$ and $m\angle 4 < m\angle 1$.

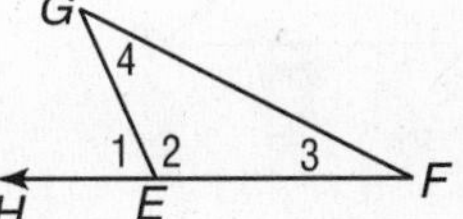

List all angles that satisfy the stated condition.

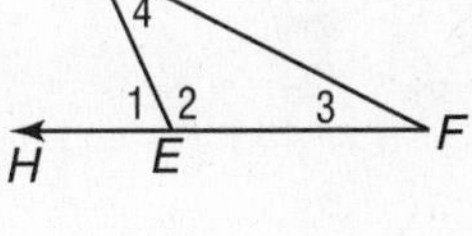

Exercises 1–2

1. all angles whose measures are less than $m\angle 1$

2. all angles whose measures are greater than $m\angle 3$

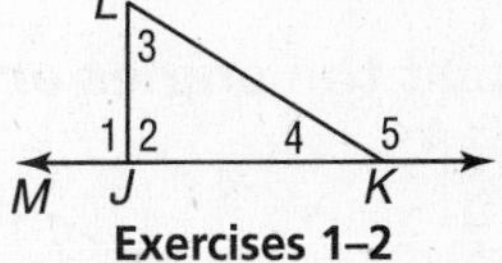

Exercises 3–8

3. all angles whose measures are less than $m\angle 1$

4. all angles whose measures are greater than $m\angle 1$

5. all angles whose measures are less than $m\angle 7$

6. all angles whose measures are greater than $m\angle 2$

7. all angles whose measures are greater than $m\angle 5$

8. all angles whose measures are less than $m\angle 4$

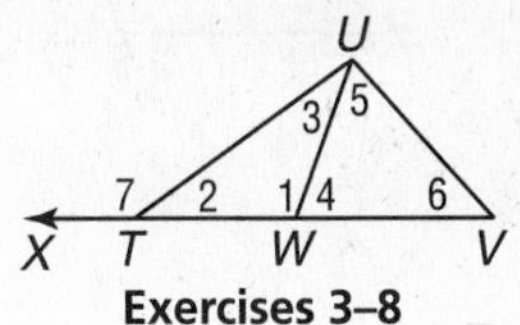

Exercises 9–10

9. all angles whose measures are less than $m\angle 1$

10. all angles whose measures are greater than $m\angle 4$

Lesson 5-2

© Glencoe/McGraw-Hill

NAME ______________________________ DATE ____________ PERIOD _____

5-2 Study Guide and Intervention *(continued)*

Inequalities and Triangles

Angle-Side Relationships

When the sides of triangles are not congruent, there is a relationship between the sides and angles of the triangles.

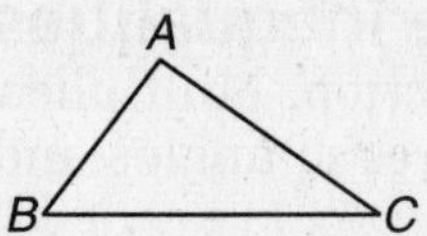

If $AC > AB$, then $m\angle B > m\angle C$.
If $m\angle A > m\angle C$, then $BC > AB$.

- If one side of a triangle is longer than another side, then the angle opposite the longer side has a greater measure than the angle opposite the shorter side.
- If one angle of a triangle has a greater measure than another angle, then the side opposite the greater angle is longer than the side opposite the lesser angle.

Example 1 **List the angles in order from least to greatest measure.**

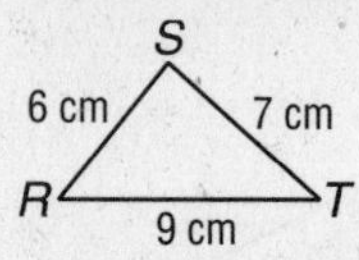

$\angle T$, $\angle R$, $\angle S$

Example 2 **List the sides in order from shortest to longest.**

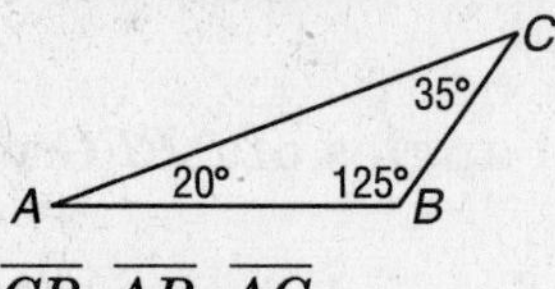

$\overline{CB}$, $\overline{AB}$, $\overline{AC}$

Exercises

List the angles or sides in order from least to greatest measure.

1.

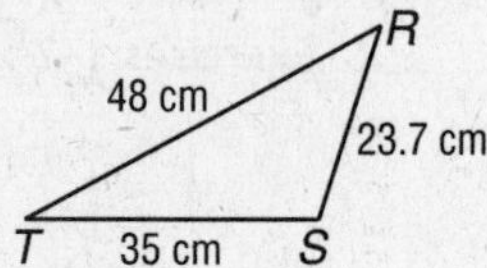

2.

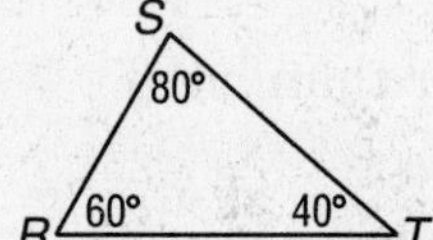

3.

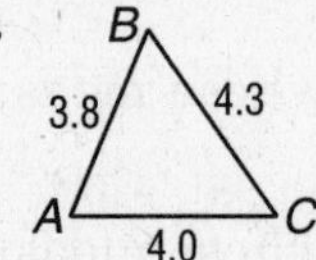

Determine the relationship between the measures of the given angles.

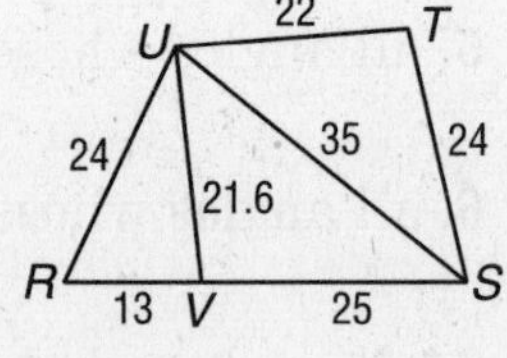

4. $\angle R$, $\angle RUS$

5. $\angle T$, $\angle UST$

6. $\angle UVS$, $\angle R$

Determine the relationship between the lengths of the given sides.

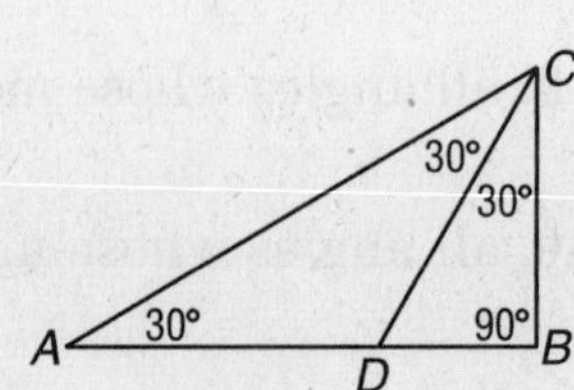

7. $\overline{AC}$, $\overline{BC}$

8. $\overline{BC}$, $\overline{DB}$

9. $\overline{AC}$, $\overline{DB}$

5-3 Study Guide and Intervention

Indirect Proof

Indirect Proof with Algebra One way to prove that a statement is true is to assume that its conclusion is false and then show that this assumption leads to a contradiction of the hypothesis, a definition, postulate, theorem, or other statement that is accepted as true. That contradiction means that the conclusion cannot be false, so the conclusion must be true. This is known as **indirect proof**.

Steps for Writing an Indirect Proof

1. Assume that the conclusion is false.
2. Show that this assumption leads to a contradiction.
3. Point out that the assumption must be false, and therefore, the conclusion must be true.

Example **Given: $3x + 5 > 8$**
Prove: $x > 1$

Step 1 Assume that x is not greater than 1. That is, $x = 1$ or $x < 1$.
Step 2 Make a table for several possibilities for $x = 1$ or $x < 1$. The contradiction is that when $x = 1$ or $x < 1$, then $3x + 5$ is not greater than 8.
Step 3 This contradicts the given information that $3x + 5 > 8$. The assumption that x is not greater than 1 must be false, which means that the statement "$x > 1$" must be true.

x	$3x + 5$
1	8
0	5
−1	2
−2	−1
−3	−4

Exercises

Write the assumption you would make to start an indirect proof of each statement.

1. If $2x > 14$, then $x > 7$.

2. For all real numbers, if $a + b > c$, then $a > c - b$.

Complete the proof.
Given: n is an integer and n^2 is even.
Prove: n is even.

3. Assume that ______________________________

4. Then n can be expressed as $2a + 1$ by ______________________________

5. n^2 = ______________ Substitution

6. = ______________ Multiply.

7. = ______________ Simplify.

8. $= 2(2a^2 + 2a) + 1$ ______________

9. $2(2a^2 + 2a) + 1$ is an odd number. This contradicts the given that n^2 is even, so the assumption must be ______________________________

10. Therefore, ______________________________

Lesson 5-3

NAME ______________________________ DATE ____________ PERIOD _____

5-3 Study Guide and Intervention *(continued)*

Indirect Proof

Indirect Proof with Geometry To write an indirect proof in geometry, you assume that the conclusion is false. Then you show that the assumption leads to a contradiction. The contradiction shows that the conclusion cannot be false, so it must be true.

Example **Given: $m\angle C = 100$**
Prove: $\angle A$ is not a right angle.

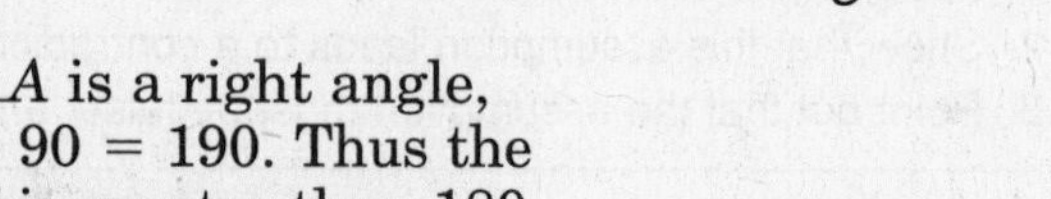

Step 1 Assume that $\angle A$ is a right angle.

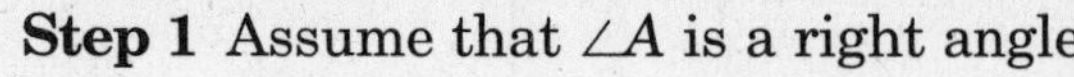

Step 2 Show that this leads to a contradiction. If $\angle A$ is a right angle, then $m\angle A = 90$ and $m\angle C + m\angle A = 100 + 90 = 190$. Thus the sum of the measures of the angles of $\triangle ABC$ is greater than 180.

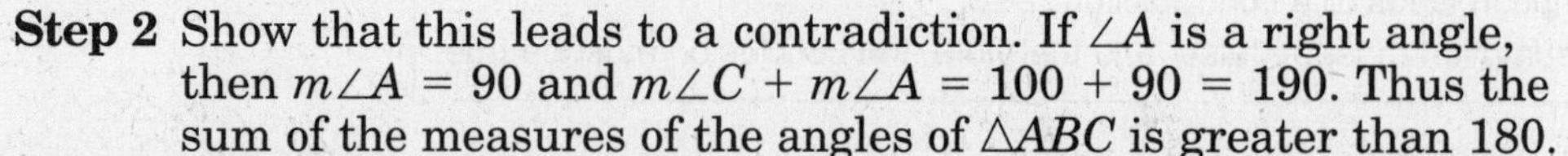

Step 3 The conclusion that the sum of the measures of the angles of $\triangle ABC$ is greater than 180 is a contradiction of a known property. The assumption that $\angle A$ is a right angle must be false, which means that the statement "$\angle A$ is not a right angle" must be true.

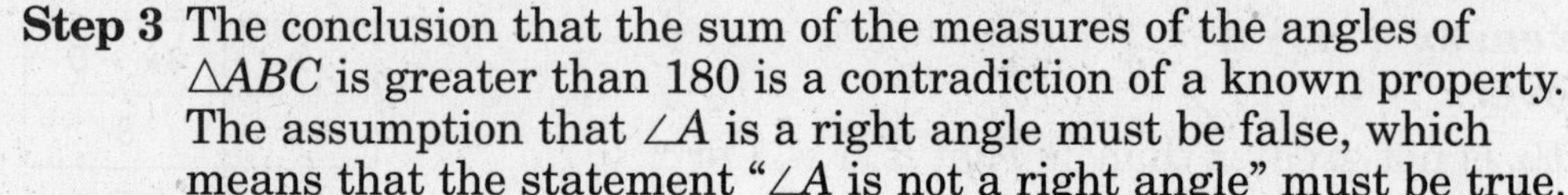

Exercises

Write the assumption you would make to start an indirect proof of each statement.

1. If $m\angle A = 90$, then $m\angle B = 45$.

2. If $\overline{AV}$ is not congruent to $\overline{VE}$, then $\triangle AVE$ is not isosceles.

Complete the proof.

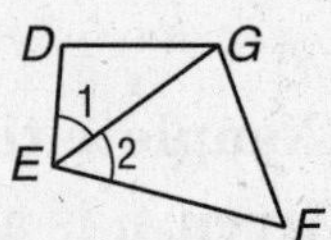

Given: $\angle 1 \cong \angle 2$ and $\overline{DG}$ is not congruent to $\overline{FG}$.
Prove: $\overline{DE}$ is not congruent to $\overline{FE}$.

3. Assume that ______________ Assume the conclusion is false.

4. $\overline{EG} \cong \overline{EG}$ ______________________________

5. $\triangle EDG \cong \triangle EFG$ ______________________________

6. ____

7. This contradicts the given information, so the assumption must be ______________________________

8. Therefore, ______________________________

NAME ______________________________ DATE __________ PERIOD _____

5-4 Study Guide and Intervention

The Triangle Inequality

The Triangle Inequality If you take three straws of lengths 8 inches, 5 inches, and 1 inch and try to make a triangle with them, you will find that it is not possible. This illustrates the Triangle Inequality Theorem.

Triangle Inequality Theorem	The sum of the lengths of any two sides of a triangle is greater than the length of the third side.	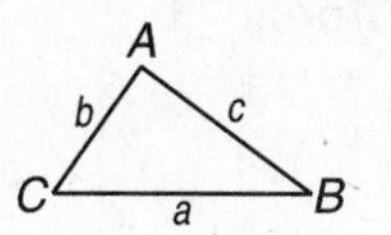

Example **The measures of two sides of a triangle are 5 and 8. Find a range for the length of the third side.**

By the Triangle Inequality, all three of the following inequalities must be true.

$5 + x > 8$	$8 + x > 5$	$5 + 8 > x$
$x > 3$	$x > -3$	$13 > x$

Therefore x must be between 3 and 13.

Exercises

Determine whether the given measures can be the lengths of the sides of a triangle. Write *yes* or *no*.

1. 3, 4, 6

2. 6, 9, 15

3. 8, 8, 8

4. 2, 4, 5

5. 4, 8, 16

6. 1.5, 2.5, 3

Find the range for the measure of the third side given the measures of two sides.

7. 1 and 6

8. 12 and 18

9. 1.5 and 5.5

10. 82 and 8

11. Suppose you have three different positive numbers arranged in order from least to greatest. What single comparison will let you see if the numbers can be the lengths of the sides of a triangle?

Lesson 5-4

NAME ______________________ DATE ____________ PERIOD ____

5-4 Study Guide and Intervention *(continued)*

The Triangle Inequality

Distance Between a Point and a Line

The perpendicular segment from a point to a line is the shortest segment from the point to the line.

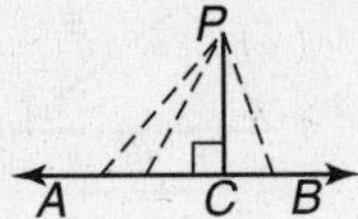

$\overline{PC}$ is the shortest segment from P to $\overleftrightarrow{AB}$.

The perpendicular segment from a point to a plane is the shortest segment from the point to the plane.

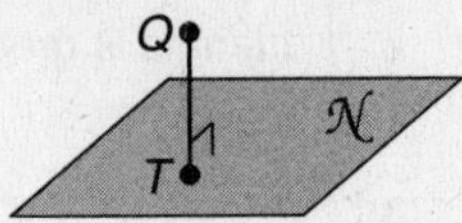

$\overline{QT}$ is the shortest segment from Q to plane $\mathcal{N}$.

Example **Given: Point P is equidistant from the sides of an angle.**
Prove: $\overline{BA} \cong \overline{CA}$

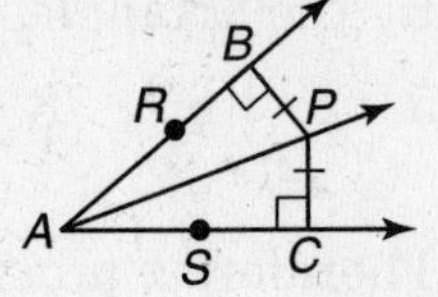

Proof:

Statements	Reasons
1. Draw $\overline{BP}$ and $\overline{CP} \perp$ to the sides of $\angle RAS$.	**1.** Dist. is measured along a $\perp$.
2. $\angle PBA$ and $\angle PCA$ are right angles.	**2.** Def. of $\perp$ lines
3. $\triangle ABP$ and $\triangle ACP$ are right triangles.	**3.** Def. of rt. $\triangle$
4. $\angle PBA \cong \angle PCA$	**4.** Rt. angles are $\cong$.
5. P is equidistant from the sides of $\angle RAS$.	**5.** Given
6. $\overline{BP} \cong \overline{CP}$	**6.** Def. of equidistant
7. $\overline{AP} \cong \overline{AP}$	**7.** Reflexive Property
8. $\triangle ABP \cong \triangle ACP$	**8.** HL
9. $\overline{BA} \cong \overline{CA}$	**9.** CPCTC

Exercises

Complete the proof.

Given: $\triangle ABC \cong \triangle RST$; $\angle D \cong \angle U$

Prove: $\overline{AD} \cong \overline{RU}$

Proof:

Statements	Reasons
1. $\triangle ABC \cong \triangle RST$; $\angle D \cong \angle U$	**1.** ____________
2. $\overline{AC} \cong \overline{RT}$	**2.** ____________
3. $\angle ACB \cong \angle RTS$	**3.** ____________
4. $\angle ACB$ and $\angle ACD$ are a linear pair; $\angle RTS$ and $\angle RTU$ are a linear pair.	**4.** Def. of ____________
5. $\angle ACB$ and $\angle ACD$ are supplementary; $\angle RTS$ and $\angle RTU$ are supplementary.	**5.** ____________
6. ____________	**6.** Angles suppl. to $\cong$ angles are $\cong$.
7. $\triangle ADC \cong \triangle RUT$	**7.** ____________
8. ____________	**8.** CPCTC

5-5 Study Guide and Intervention

Inequalities Involving Two Triangles

SAS Inequality The following theorem involves the relationship between the sides of two triangles and an angle in each triangle.

SAS Inequality/Hinge Theorem	If two sides of a triangle are congruent to two sides of another triangle and the included angle in one triangle has a greater measure than the included angle in the other, then the third side of the first triangle is longer than the third side of the second triangle.	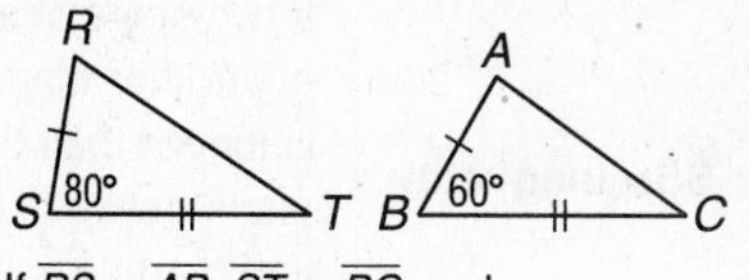 If $\overline{RS} \cong \overline{AB}$, $\overline{ST} \cong \overline{BC}$, and $m\angle S > m\angle B$, then $RT > AC$.

Example **Write an inequality relating the lengths of $\overline{CD}$ and $\overline{AD}$.**

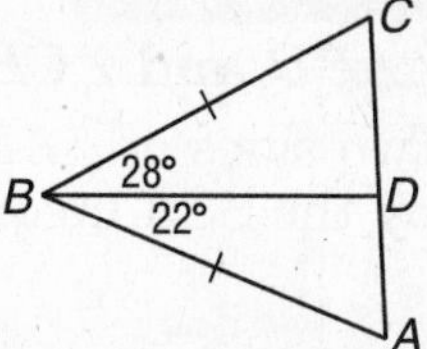

Two sides of $\triangle BCD$ are congruent to two sides of $\triangle BAD$ and $m\angle CBD > m\angle ABD$. By the SAS Inequality/Hinge Theorem, $CD > AD$.

Exercises

Write an inequality relating the given pair of segment measures.

1. 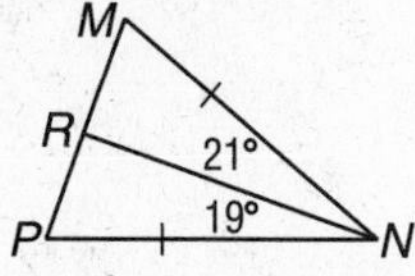

MR, RP

2. 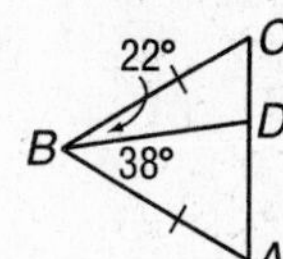

AD, CD

3. 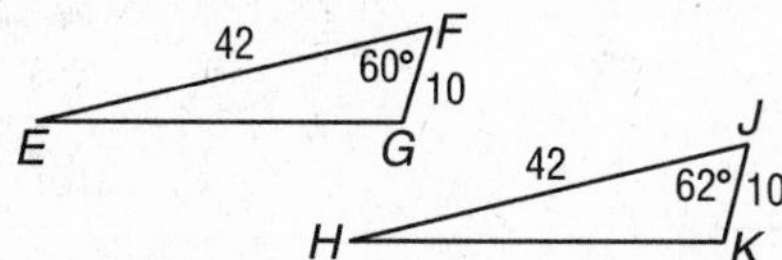

EG, HK

4. 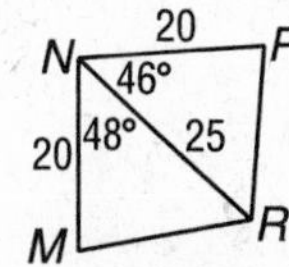

MR, PR

Write an inequality to describe the possible values of x.

5. 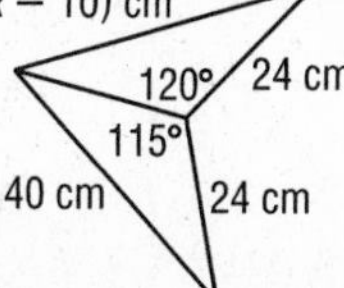

6. 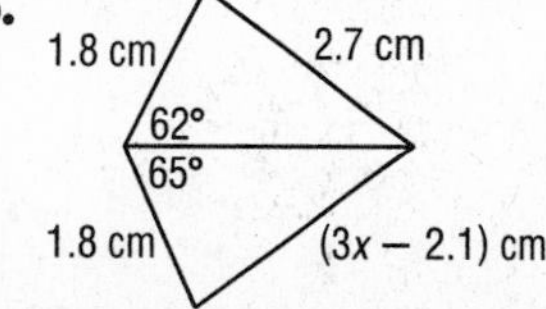

Study Guide and Intervention *(continued)*

Inequalities Involving Two Triangles

SSS Inequality The converse of the Hinge Theorem is also useful when two triangles have two pairs of congruent sides.

SSS Inequality	If two sides of a triangle are congruent to two sides of another triangle and the third side in one triangle is longer than the third side in the other, then the angle between the pair of congruent sides in the first triangle is greater than the corresponding angle in the second triangle.	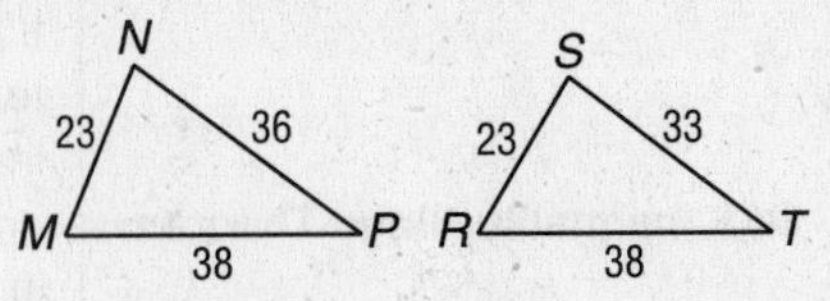 If $NM = SR$, $MP = RT$, and $NP > ST$, then $m\angle M > m\angle R$.

Example **Write an inequality relating the measures of $\angle ABD$ and $\angle CBD$.**

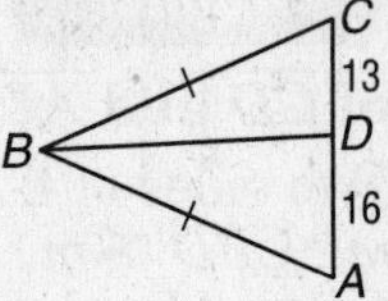

Two sides of $\triangle ABD$ are congruent to two sides of $\triangle CBD$, and $AD > CD$. By the SSS Inequality, $m\angle ABD > m\angle CBD$.

Exercises

Write an inequality relating the given pair of angle measures.

1.

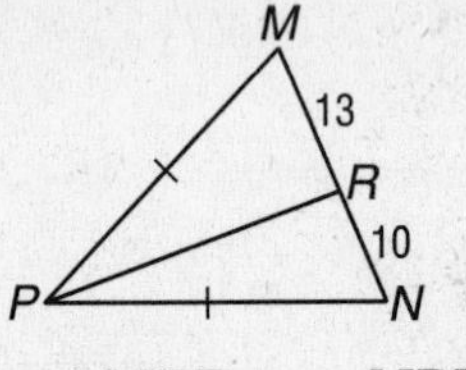

$m\angle MPR$, $m\angle NPR$

2.

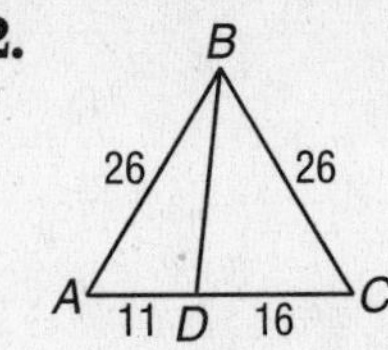

$m\angle ABD$, $m\angle CBD$

3.

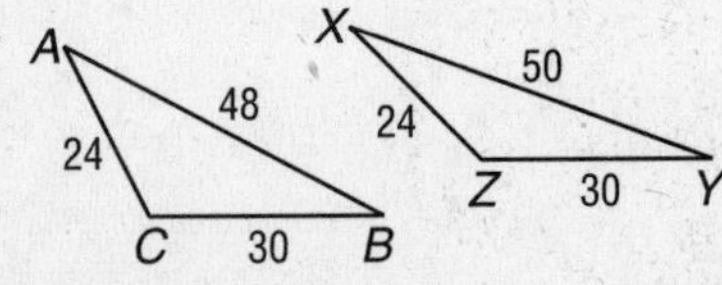

$m\angle C$, $m\angle Z$

4.

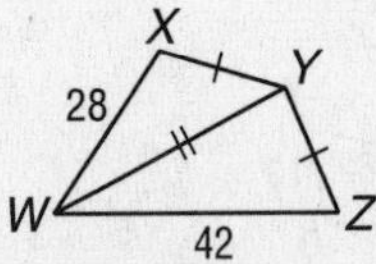

$m\angle XYW$, $m\angle WYZ$

Write an inequality to describe the possible values of x.

5.

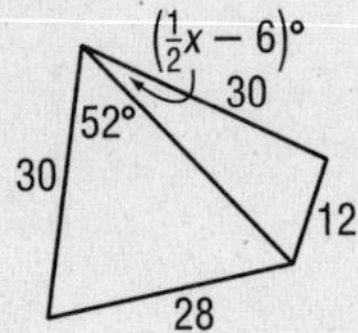

6.

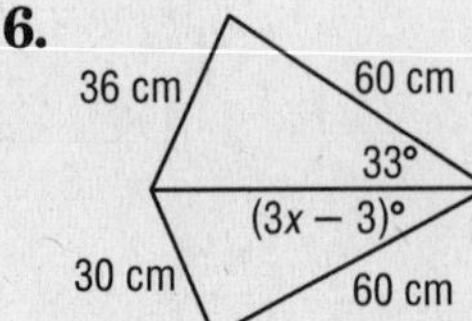

NAME ______________________ DATE ____________ PERIOD _____

6-1 Study Guide and Intervention

Proportions

Write Ratios A **ratio** is a comparison of two quantities. The ratio a to b, where b is not zero, can be written as $\frac{a}{b}$ or $a{:}b$. The ratio of two quantities is sometimes called a **scale factor**. For a scale factor, the units for each quantity are the same.

Example 1 **In 2002, the Chicago Cubs baseball team won 67 games out of 162. Write a ratio for the number of games won to the total number of games played.**

To find the ratio, divide the number of games won by the total number of games played. The result is $\frac{67}{162}$, which is about 0.41. The Chicago Cubs won about 41% of their games in 2002.

Example 2 **A doll house that is 15 inches tall is a scale model of a real house with a height of 20 feet. What is the ratio of the height of the doll house to the height of the real house?**

To start, convert the height of the real house to inches.

20 feet $\times$ 12 inches per foot $=$ 240 inches

To find the ratio or scale factor of the heights, divide the height of the doll house by the height of the real house. The ratio is 15 inches:240 inches or 1:16. The height of the doll house is $\frac{1}{16}$ the height of the real house.

Exercises

1. In the 2002 Major League baseball season, Sammy Sosa hit 49 home runs and was at bat 556 times. Find the ratio of home runs to the number of times he was at bat.

2. There are 182 girls in the sophomore class of 305 students. Find the ratio of girls to total students.

3. The length of a rectangle is 8 inches and its width is 5 inches. Find the ratio of length to width.

4. The sides of a triangle are 3 inches, 4 inches, and 5 inches. Find the scale factor between the longest and the shortest sides.

5. The length of a model train is 18 inches. It is a scale model of a train that is 48 feet long. Find the scale factor.

6-1 Study Guide and Intervention *(continued)*

Proportions

Use Properties of Proportions A statement that two ratios are equal is called a **proportion**. In the proportion $\frac{a}{b} = \frac{c}{d}$, where b and d are not zero, the values a and d are the **extremes** and the values b and c are the **means**. In a proportion, the product of the means is equal to the product of the extremes, so $ad = bc$.

$$\frac{a}{b} = \frac{c}{d}$$

$$a \cdot d = b \cdot c$$

↑ extremes ↑ means

Example 1 **Solve $\frac{9}{16} = \frac{27}{x}$.**

$\frac{9}{16} = \frac{27}{x}$

$9 \cdot x = 16 \cdot 27$ Cross products

$9x = 432$ Multiply.

$x = 48$ Divide each side by 9.

Example 2 **A room is 49 centimeters by 28 centimeters on a scale drawing of a house. For the actual room, the larger dimension is 14 feet. Find the shorter dimension of the actual room.**

If x is the room's shorter dimension, then

$\frac{28}{49} = \frac{x}{14}$ $\frac{\text{shorter dimension}}{\text{longer dimension}}$

$49x = 392$ Cross products

$x = 8$ Divide each side by 49.

The shorter side of the room is 8 feet.

Exercises

Solve each proportion.

1. $\frac{1}{2} = \frac{28}{x}$

2. $\frac{3}{8} = \frac{y}{24}$

3. $\frac{x + 22}{x + 2} = \frac{30}{10}$

4. $\frac{3}{18.2} = \frac{9}{y}$

5. $\frac{2x + 3}{8} = \frac{5}{4}$

6. $\frac{x + 1}{x - 1} = \frac{3}{4}$

Use a proportion to solve each problem.

7. If 3 cassettes cost $44.85, find the cost of one cassette.

8. The ratio of the sides of a triangle are 8:15:17. If the perimeter of the triangle is 480 inches, find the length of each side of the triangle.

9. The scale on a map indicates that one inch equals 4 miles. If two towns are 3.5 inches apart on the map, what is the actual distance between the towns?

6-2 Study Guide and Intervention

Similar Polygons

Identify Similar Figures

Example 1 **Determine whether the triangles are similar.**

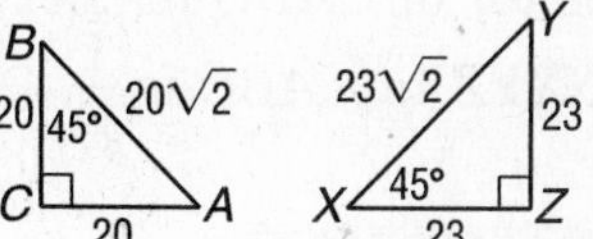

Two polygons are similar if and only if their corresponding angles are congruent and their corresponding sides are proportional.

$\angle C \cong \angle Z$ because they are right angles, and $\angle B \cong \angle X$. By the Third Angle Theorem, $\angle A \cong \angle Y$.

For the sides, $\frac{BC}{XZ} = \frac{20}{23}$, $\frac{BA}{XY} = \frac{20\sqrt{2}}{23\sqrt{2}} = \frac{20}{23}$, and $\frac{AC}{YZ} = \frac{20}{23}$.

The side lengths are proportional. So $\triangle BCA \sim \triangle XZY$.

Example 2 **Is polygon *WXYZ* ~ polygon *PQRS*?**

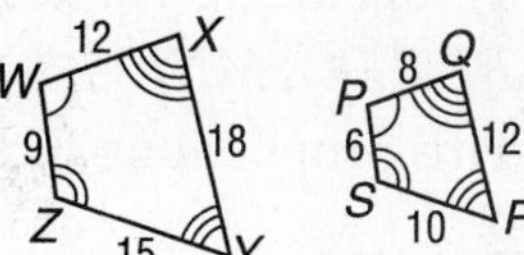

For the sides, $\frac{WX}{PQ} = \frac{12}{8} = \frac{3}{2}$, $\frac{XY}{QR} = \frac{18}{12} = \frac{3}{2}$, $\frac{YZ}{RS} = \frac{15}{10} = \frac{3}{2}$, and $\frac{ZW}{SP} = \frac{9}{6} = \frac{3}{2}$. So corresponding sides are proportional.

Also, $\angle W \cong \angle P$, $\angle X \cong \angle Q$, $\angle Y \cong \angle R$, and $\angle Z \cong \angle S$, so corresponding angles are congruent. We can conclude that polygon $WXYZ \sim$ polygon $PQRS$.

Exercises

Determine whether each pair of figures is similar. If they are similar, give the ratio of corresponding sides.

1. 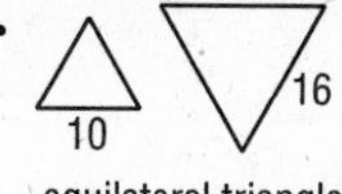

2. 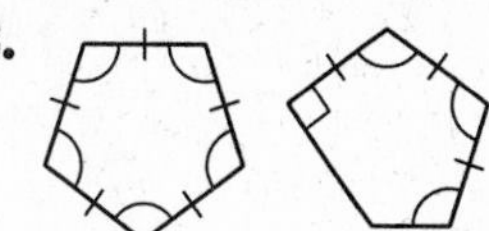

3. 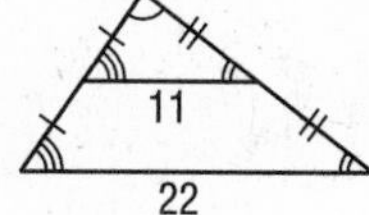

4. 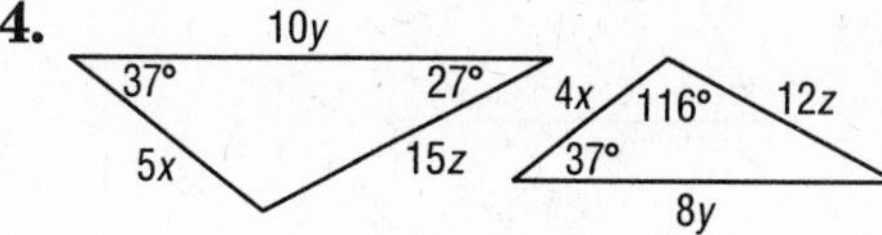

Lesson 6-2

NAME ______________________ DATE ____________ PERIOD ______

6-2 Study Guide and Intervention *(continued)*

Similar Polygons

Scale Factors When two polygons are similar, the ratio of the lengths of corresponding sides is called the **scale factor**. At the right, $\triangle ABC \sim \triangle XYZ$. The scale factor of $\triangle ABC$ to $\triangle XYZ$ is 2 and the scale factor of $\triangle XYZ$ to $\triangle ABC$ is $\frac{1}{2}$.

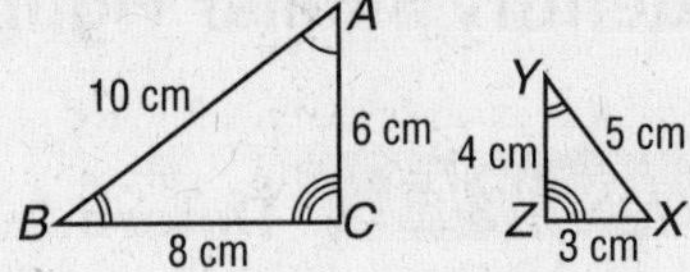

Example 1 **The two polygons are similar. Find x and y.**

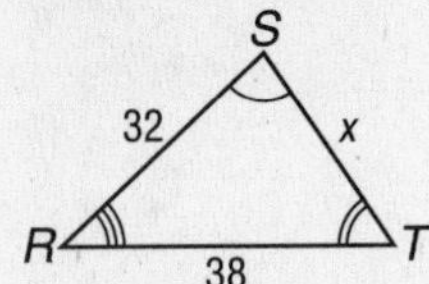

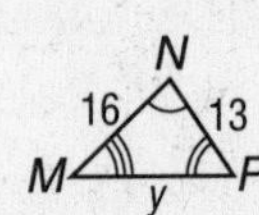

Use the congruent angles to write the corresponding vertices in order.

$\triangle RST \sim \triangle MNP$

Write proportions to find x and y.

$\frac{32}{16} = \frac{x}{13}$ $\qquad$ $\frac{38}{y} = \frac{32}{16}$

$16x = 32(13)$ $\qquad$ $32y = 38(16)$

$x = 26$ $\qquad$ $y = 19$

Example 2 **$\triangle ABC \sim \triangle CDE$. Find the scale factor and find the lengths of $\overline{CD}$ and $\overline{DE}$.**

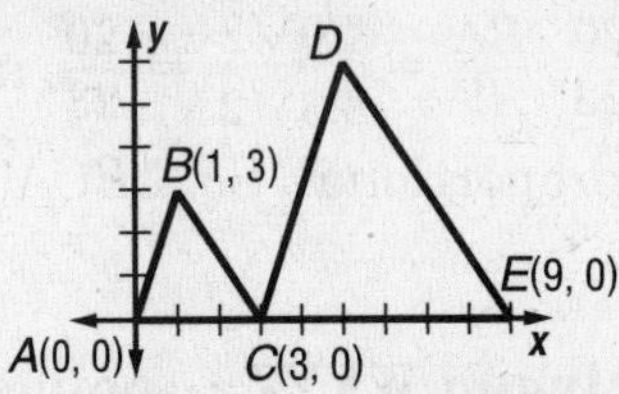

$AC = 3 - 0 = 3$ and $CE = 9 - 3 = 6$. The scale factor of $\triangle CDE$ to $\triangle ABC$ is 6:3 or 2:1. Using the Distance Formula, $AB = \sqrt{1 + 9} = \sqrt{10}$ and $BC = \sqrt{4 + 9} = \sqrt{13}$. The lengths of the sides of $\triangle CDE$ are twice those of $\triangle ABC$, so $DC = 2(BA)$ or $2\sqrt{10}$ and $DE = 2(BC)$ or $2\sqrt{13}$.

Exercises

Each pair of polygons is similar. Find x and y.

1.

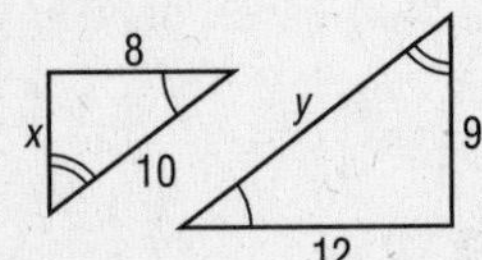

2.

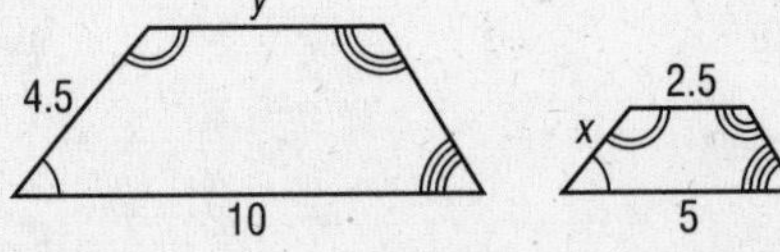

3.

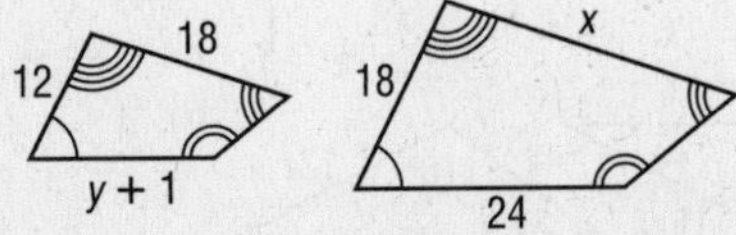

4.

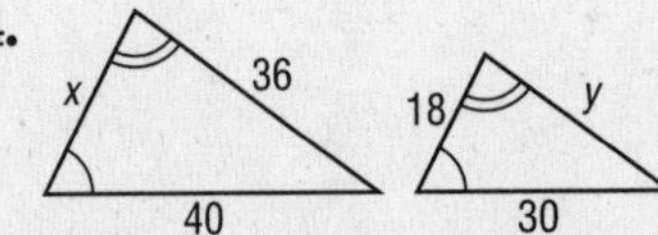

5. In Example 2 above, point D has coordinates (5, 6). Use the Distance Formula to verify the lengths of $\overline{CD}$ and $\overline{DE}$.

NAME ______________________ DATE ____________ PERIOD _____

6-3 Study Guide and Intervention

Similar Triangles

Identify Similar Triangles Here are three ways to show that two triangles are similar.

AA Similarity	Two angles of one triangle are congruent to two angles of another triangle.
SSS Similarity	The measures of the corresponding sides of two triangles are proportional.
SAS Similarity	The measures of two sides of one triangle are proportional to the measures of two corresponding sides of another triangle, and the included angles are congruent.

Example 1 **Determine whether the triangles are similar.**

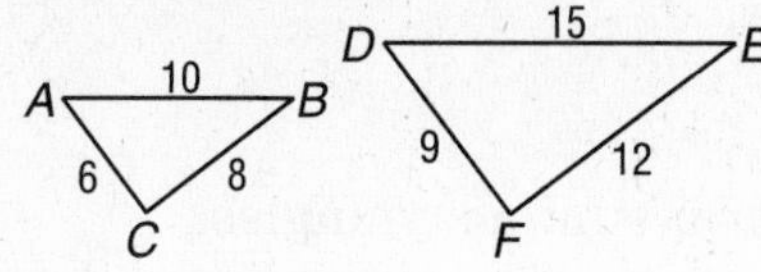

$\frac{AC}{DF} = \frac{6}{9} = \frac{2}{3}$

$\frac{BC}{EF} = \frac{8}{12} = \frac{2}{3}$

$\frac{AB}{DE} = \frac{10}{15} = \frac{2}{3}$

$\triangle ABC \sim \triangle DEF$ by SSS Similarity.

Example 2 **Determine whether the triangles are similar.**

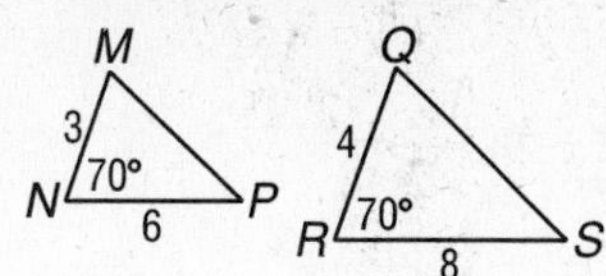

$\frac{3}{4} = \frac{6}{8}$, so $\frac{MN}{QR} = \frac{NP}{RS}$.

$m\angle N = m\angle R$, so $\angle N \cong \angle R$.

$\triangle NMP \sim \triangle RQS$ by SAS Similarity.

Exercises

Determine whether each pair of triangles is similar. Justify your answer.

1. 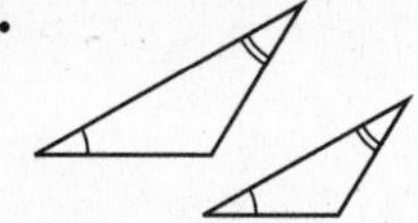

2. 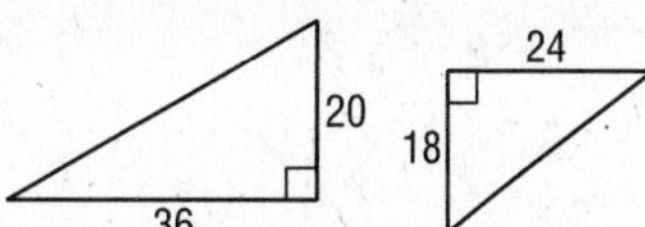

3. 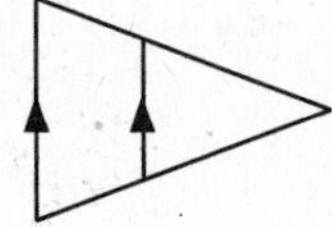

4. 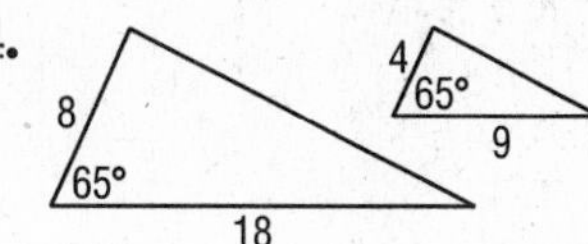

5. 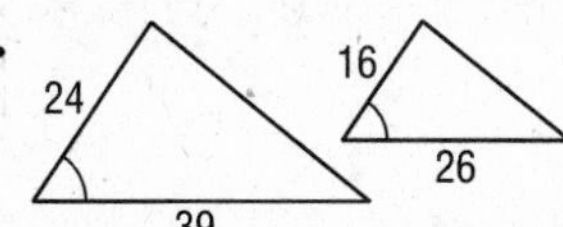

6. 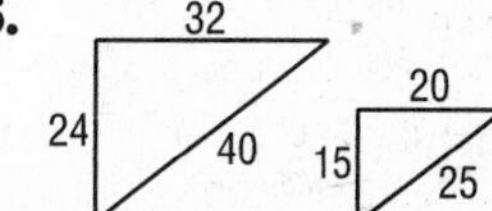

Lesson 6-3

NAME ______________________________ DATE ____________ PERIOD _____

6-3 Study Guide and Intervention *(continued)*

Similar Triangles

Use Similar Triangles

Similar triangles can be used to find measurements.

Example 1 $\triangle ABC \sim \triangle DEF$.
Find x and y.

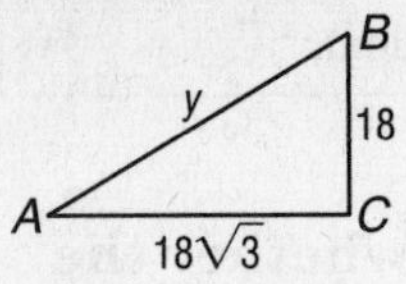

$$\frac{AC}{DF} = \frac{BC}{EF} \qquad \frac{AB}{DE} = \frac{BC}{EF}$$

$$\frac{18\sqrt{3}}{x} = \frac{18}{9} \qquad \frac{y}{18} = \frac{18}{9}$$

$$18x = 9(18\sqrt{3}) \qquad 9y = 324$$

$$x = 9\sqrt{3} \qquad y = 36$$

Example 2 **A person 6 feet tall casts a 1.5-foot-long shadow at the same time that a flagpole casts a 7-foot-long shadow. How tall is the flagpole?**

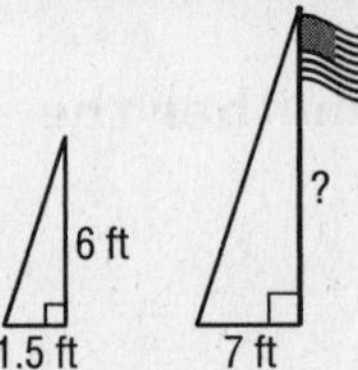

The sun's rays form similar triangles.
Using x for the height of the pole, $\frac{6}{x} = \frac{1.5}{7}$,
so $1.5x = 42$ and $x = 28$.
The flagpole is 28 feet tall.

Exercises

Each pair of triangles is similar. Find x and y.

1.

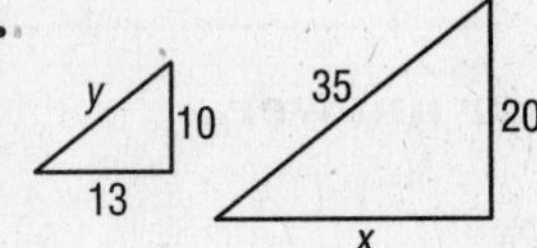

2.

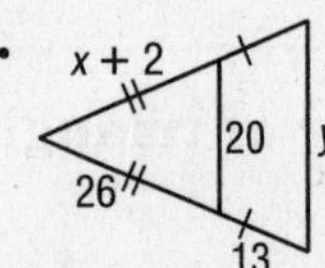

3.

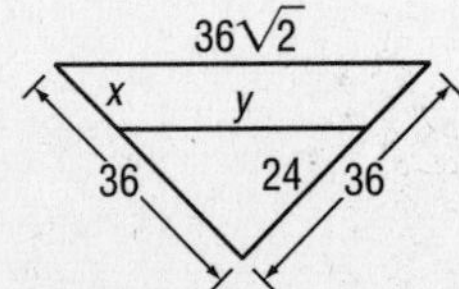

4.

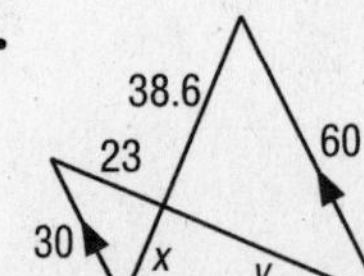

5.

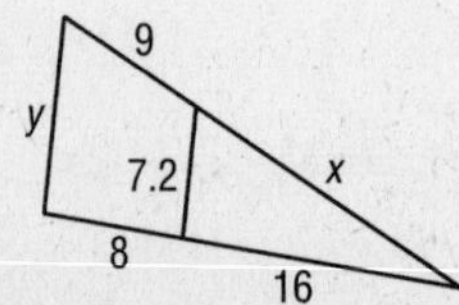

6.

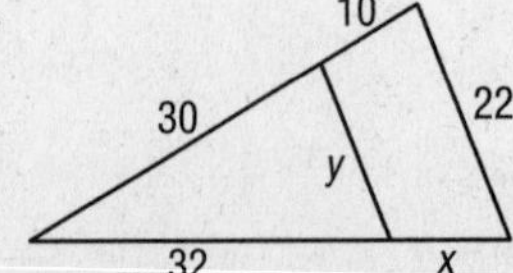

7. The heights of two vertical posts are 2 meters and 0.45 meter. When the shorter post casts a shadow that is 0.85 meter long, what is the length of the longer post's shadow to the nearest hundredth?

6-4 Study Guide and Intervention

Parallel Lines and Proportional Parts

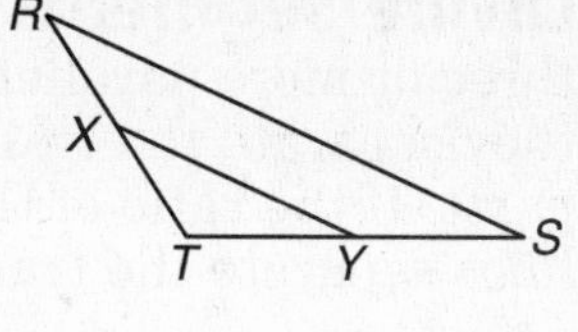

Proportional Parts of Triangles In any triangle, a line parallel to one side of a triangle separates the other two sides proportionally. The converse is also true.

If X and Y are the midpoints of $\overline{RT}$ and $\overline{ST}$, then $\overline{XY}$ is a **midsegment** of the triangle. The Triangle Midsegment Theorem states that a midsegment is parallel to the third side and is half its length.

If $\overleftrightarrow{XY} \parallel \overleftrightarrow{RS}$, then $\frac{RX}{XT} = \frac{SY}{YT}$.

If $\frac{RX}{XT} = \frac{SY}{YT}$, then $\overleftrightarrow{XY} \parallel \overleftrightarrow{RS}$.

If $\overline{XY}$ is a midsegment, then $\overleftrightarrow{XY} \parallel \overleftrightarrow{RS}$ and $XY = \frac{1}{2}RS$.

Example 1 **In $\triangle ABC$, $\overline{EF} \parallel \overline{CB}$. Find x.**

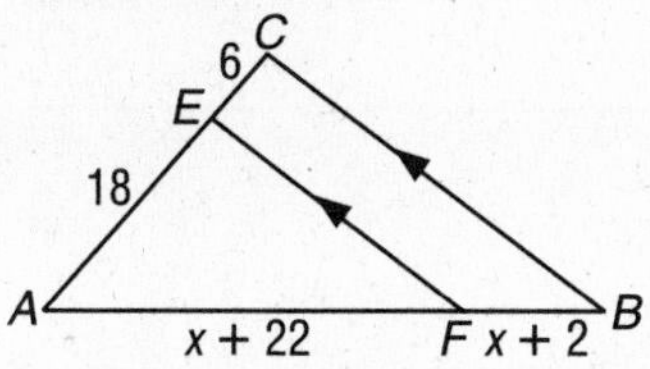

Since $\overline{EF} \parallel \overline{CB}$, $\frac{AF}{FB} = \frac{AE}{EC}$.

$$\frac{x + 22}{x + 2} = \frac{18}{6}$$
$$6x + 132 = 18x + 36$$
$$96 = 12x$$
$$8 = x$$

Example 2 **A triangle has vertices $D(3, 6)$, $E(-3, -2)$, and $F(7, -2)$. Midsegment $\overline{GH}$ is parallel to $\overline{EF}$. Find the length of $\overline{GH}$.**

$\overline{GH}$ is a midsegment, so its length is one-half that of $\overline{EF}$. Points E and F have the same y-coordinate, so $EF = 7 - (-3) = 10$. The length of midsegment $\overline{GH}$ is 5.

Exercises

Find x.

1.
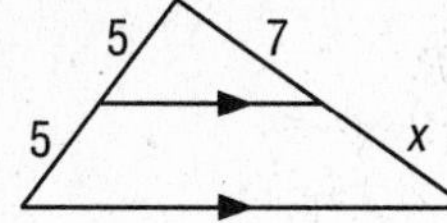

2.
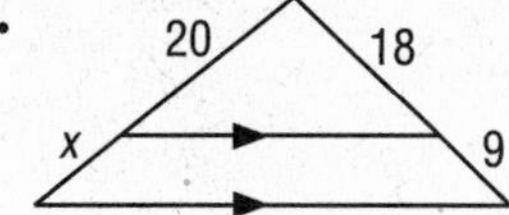

3.
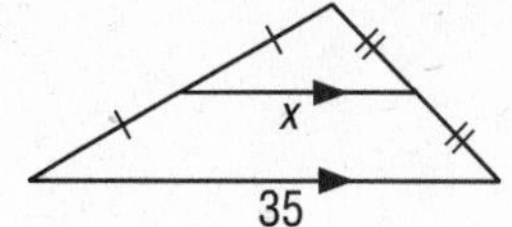

4.
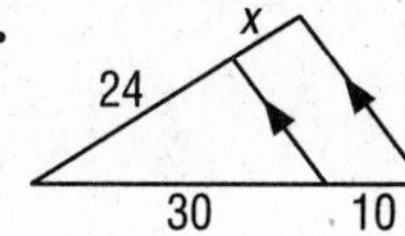

5.
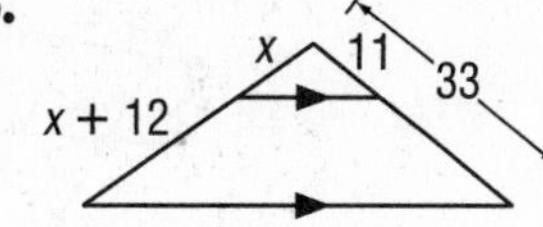

6.
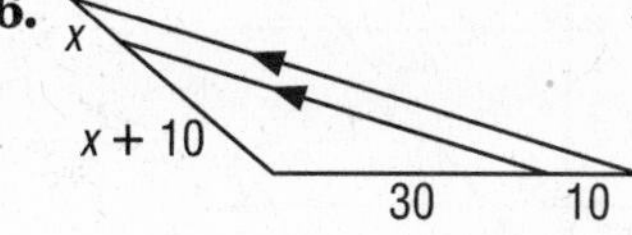

7. In Example 2, find the slope of $\overline{EF}$ and show that $\overline{EF} \parallel \overline{GH}$.

Lesson 6-4

NAME ______________________________ DATE ____________ PERIOD _____

6-4 Study Guide and Intervention *(continued)*

Parallel Lines and Proportional Parts

Divide Segments Proportionally When three or more parallel lines cut two transversals, they separate the transversals into proportional parts. If the ratio of the parts is 1, then the parallel lines separate the transversals into congruent parts.

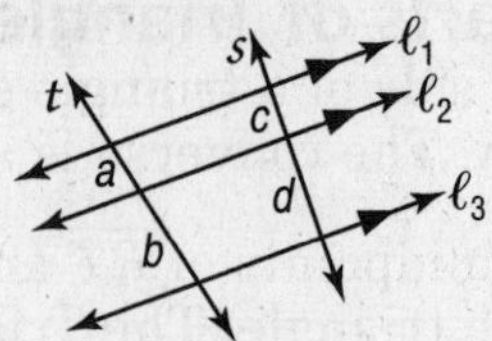

If $\ell_1 \parallel \ell_2 \parallel \ell_3$, then $\frac{a}{b} = \frac{c}{d}$.

If $\ell_4 \parallel \ell_5 \parallel \ell_6$ and $\frac{u}{v} = 1$, then $\frac{w}{x} = 1$.

Example **Refer to lines ℓ_1, ℓ_2, and ℓ_3 above. If $a = 3$, $b = 8$, and $c = 5$, find d.**

$\ell_1 \parallel \ell_2 \parallel \ell_3$ so $\frac{3}{8} = \frac{5}{d}$. Then $3d = 40$ and $d = 13\frac{1}{3}$.

Exercises

Find x and y.

1.

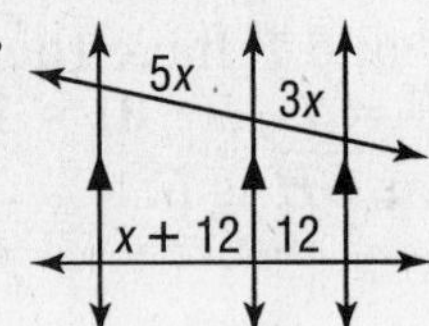

2.

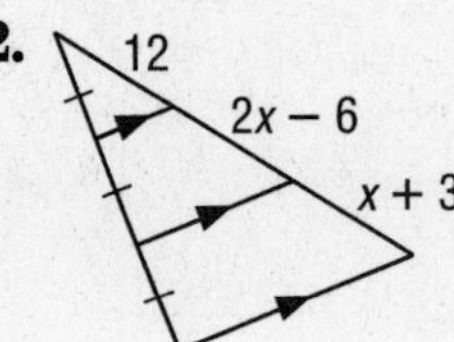

3.

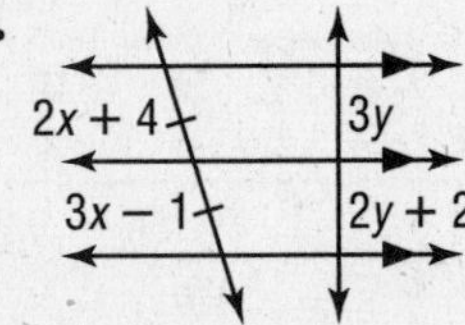

4.

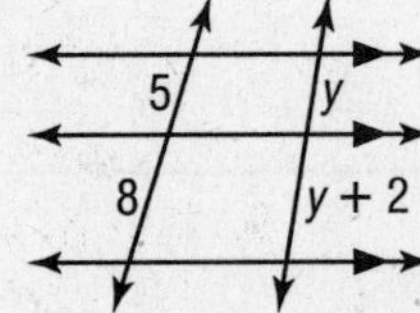

5.

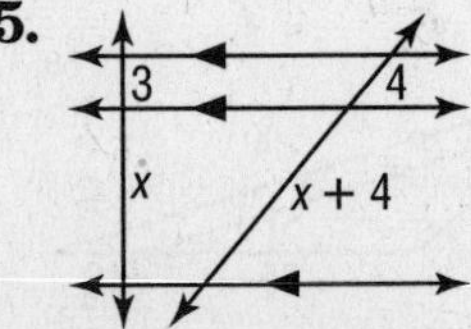

6.

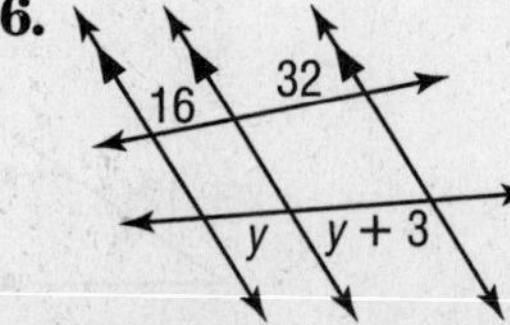

6-5 Study Guide and Intervention

Parts of Similar Triangles

Perimeters If two triangles are similar, their perimeters have the same proportion as the corresponding sides.

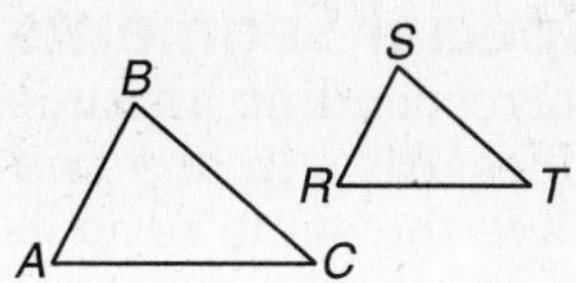

If $\triangle ABC \sim \triangle RST$, then

$$\frac{AB + BC + AC}{RS + ST + RT} = \frac{AB}{RS} = \frac{BC}{ST} = \frac{AC}{RT}.$$

Example **Use the diagram above with $\triangle ABC \sim \triangle RST$. If $AB = 24$ and $RS = 15$, find the ratio of their perimeters.**

Since $\triangle ABC \sim \triangle RST$, the ratio of the perimeters of $\triangle ABC$ and $\triangle RST$ is the same as the ratio of corresponding sides.

Therefore $\frac{\text{perimeter of } \triangle ABC}{\text{perimeter of } \triangle RST} = \frac{24}{15}$

$= \frac{8}{5}$

Exercises

Each pair of triangles is similar. Find the perimeter of the indicated triangle.

1. $\triangle XYZ$

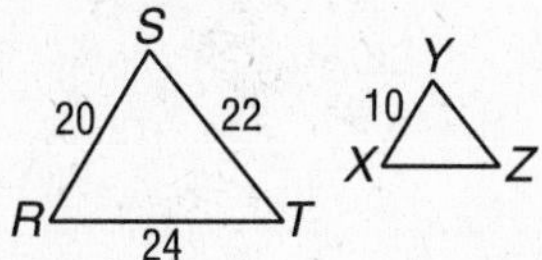

2. $\triangle BDE$

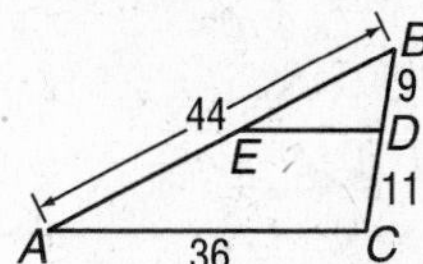

3. $\triangle ABC$

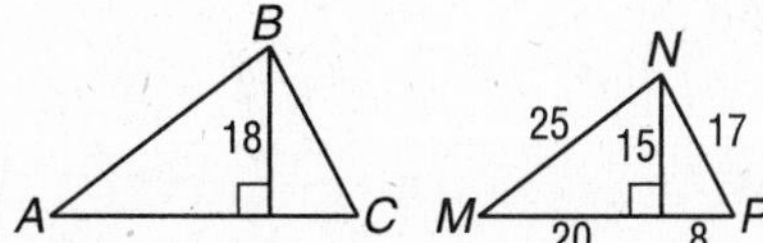

4. $\triangle XYZ$

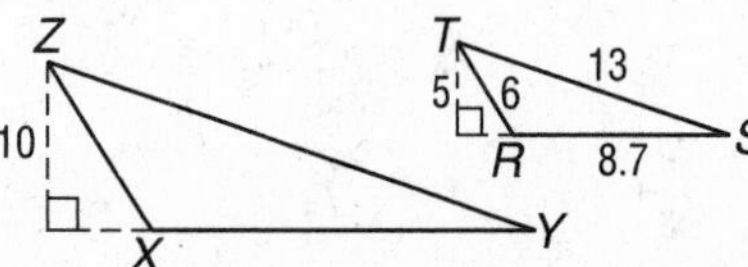

5. $\triangle ABC$

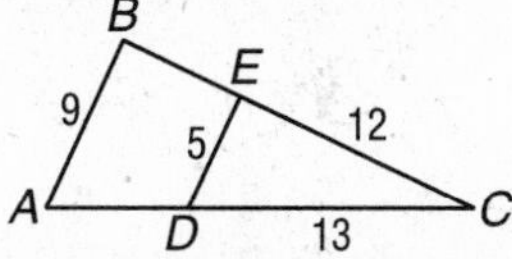

6. $\triangle RST$

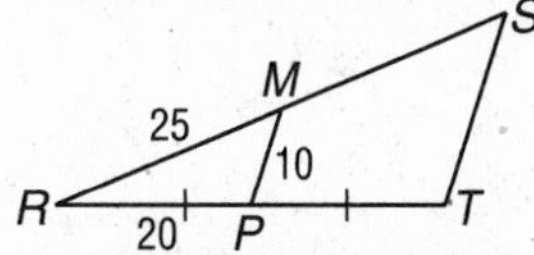

Lesson 6-5

6-5 Study Guide and Intervention *(continued)*

Parts of Similar Triangles

Special Segments of Similar Triangles When two triangles are similar, corresponding altitudes, angle bisectors, and medians are proportional to the corresponding sides. Also, in any triangle an angle bisector separates the opposite side into segments that have the same ratio as the other two sides of the triangle.

Example 1 **In the figure, $\triangle ABC \sim \triangle XYZ$, with angle bisectors as shown. Find x.**

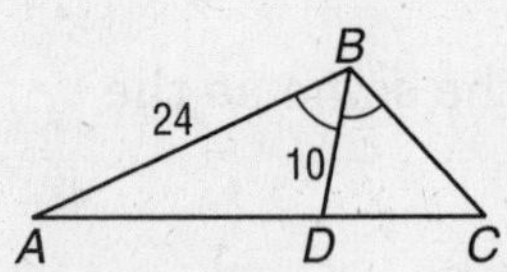

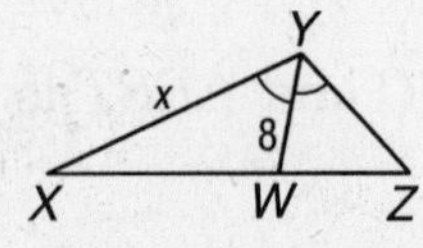

Since $\triangle ABC \sim \triangle XYZ$, the measures of the angle bisectors are proportional to the measures of a pair of corresponding sides.

$$\frac{AB}{XY} = \frac{BD}{YW}$$

$$\frac{24}{x} = \frac{10}{8}$$

$$10x = 24(8)$$

$$10x = 192$$

$$x = 19.2$$

Example 2 **$\overline{SU}$ bisects $\angle RST$. Find x.**

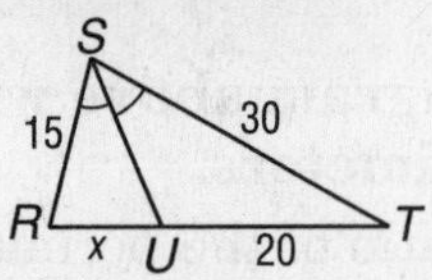

Since $\overline{SU}$ is an angle bisector, $\frac{RU}{TU} = \frac{RS}{TS}$.

$$\frac{x}{20} = \frac{15}{30}$$

$$30x = 20(15)$$

$$30x = 300$$

$$x = 10$$

Exercises

Find x for each pair of similar triangles.

1.

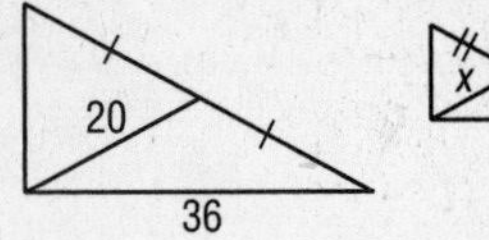

2.

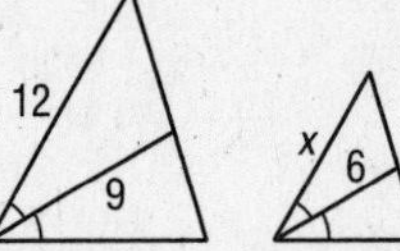

3.

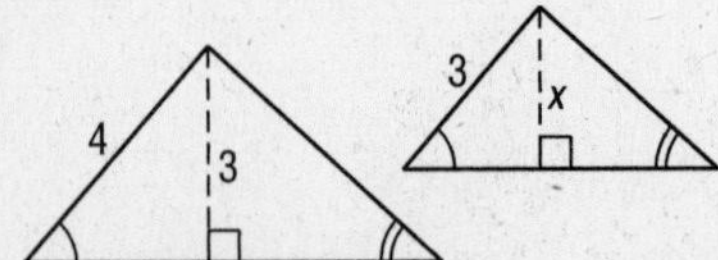

4.

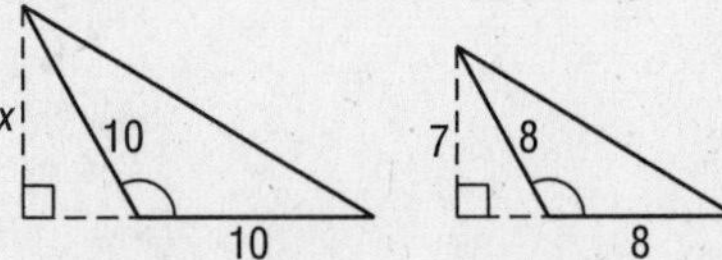

5.

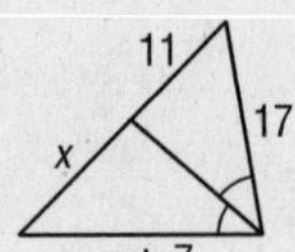

6.

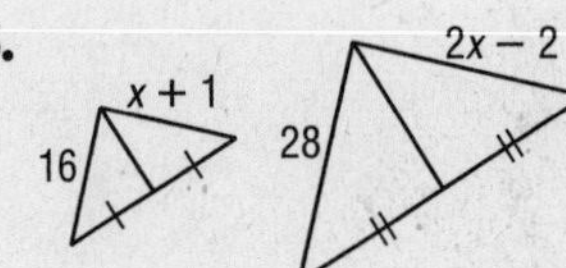

NAME ______________________ DATE ____________ PERIOD _____

6-6 Study Guide and Intervention

Fractals and Self-Similarity

Lesson 6-6

Characteristics of Fractals The act of repeating a process over and over, such as finding a third of a segment, then a third of the new segment, and so on, is called **iteration**. When the process of iteration is applied to some geometric figures, the results are called **fractals**. For objects such as fractals, when a portion of the object has the same shape or characteristics as the entire object, the object can be called **self-similar**.

Example **In the diagram at the right, notice that the details at each stage are similar to the details at Stage 1.**

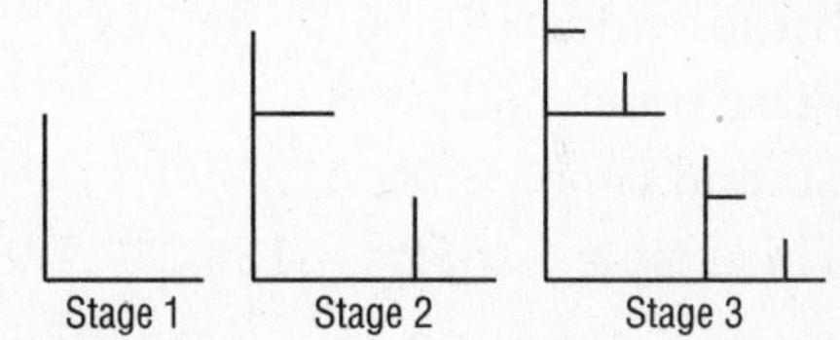

Exercises

1. Follow the iteration process below to produce a fractal.

Stage 1

- Draw a square.
- Draw an isosceles right triangle on the top side of the square. Use the side of the square as the hypotenuse of the triangle.
- Draw a square on each leg of the right triangle.

Stage 2

Repeat the steps in Stage 1, drawing an isosceles triangle and two small squares for each of the small squares from Stage 1.

Stage 3

Repeat the steps in Stage 1 for each of the smallest squares in Stage 2.

2. Is the figure produced in Stage 3 self-similar?

© Glencoe/McGraw-Hill

6-6 Study Guide and Intervention *(continued)*

Fractals and Self-Similarity

Nongeometric Iteration An iterative process can be applied to an algebraic expression or equation. The result is called a **recursive formula.**

Example **Find the value of x^3, where the initial value of x is 2. Repeat the process three times and describe the pattern.**

Initial value: 2
First time: $2^3 = 8$
Second time: $8^3 = 512$
Third time: $512^3 = 134{,}217{,}728$

The result of each step of the iteration is used for the next step. For this example, the x values are greater with each iteration. There is no maximum value, so the values are described as *approaching infinity*.

Exercises

For Exercises 1–5, find the value of each expression. Then use that value as the next x in the expression. Repeat the process three times, and describe your observations.

1. $\sqrt{x}$, where x initially equals 5

2. $\frac{1}{x}$, where x initially equals 2

3. 3^x, where x initially equals 1

4. $x - 5$ where x initially equals 10

5. $x^2 - 4$, where x initially equals 16

6. Harpesh paid $1000 for a savings certificate. It earns interest at an annual rate of 2.8%, and interest is added to the certificate each year. What will the certificate be worth after four years?

NAME ______________________ DATE __________ PERIOD _____

7-1 Study Guide and Intervention

Geometric Mean

Geometric Mean The **geometric mean** between two numbers is the square root of their product. For two positive numbers a and b, the geometric mean of a and b is the positive number x in the proportion $\frac{a}{x} = \frac{x}{b}$. Cross multiplying gives $x^2 = ab$, so $x = \sqrt{ab}$.

Example **Find the geometric mean between each pair of numbers.**

a. 12 and 3

Let x represent the geometric mean.

$\frac{12}{x} = \frac{x}{3}$ Definition of geometric mean

$x^2 = 36$ Cross multiply.

$x = \sqrt{36}$ or 6 Take the square root of each side.

b. 8 and 4

Let x represent the geometric mean.

$\frac{8}{x} = \frac{x}{4}$

$x^2 = 32$

$x = \sqrt{32}$

≈ 5.7

Exercises

Find the geometric mean between each pair of numbers.

1. 4 and 4

2. 4 and 6

3. 6 and 9

4. $\frac{1}{2}$ and 2

5. $2\sqrt{3}$ and $3\sqrt{3}$

6. 4 and 25

7. $\sqrt{3}$ and $\sqrt{6}$

8. 10 and 100

9. $\frac{1}{2}$ and $\frac{1}{4}$

10. $\frac{2\sqrt{2}}{5}$ and $\frac{3\sqrt{2}}{5}$

11. 4 and 16

12. 3 and 24

The geometric mean and one extreme are given. Find the other extreme.

13. $\sqrt{24}$ is the geometric mean between a and b. Find b if $a = 2$.

14. $\sqrt{12}$ is the geometric mean between a and b. Find b if $a = 3$.

Determine whether each statement is *always, sometimes,* or *never* true.

15. The geometric mean of two positive numbers is greater than the average of the two numbers.

16. If the geometric mean of two positive numbers is less than 1, then both of the numbers are less than 1.

Lesson 7-1

7-1 Study Guide and Intervention *(continued)*

Geometric Mean

Altitude of a Triangle In the diagram, $\triangle ABC \sim \triangle ADB \sim \triangle BDC$. An altitude to the hypotenuse of a right triangle forms two right triangles. The two triangles are similar and each is similar to the original triangle.

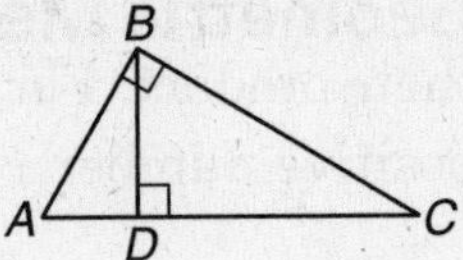

Example 1 **Use right $\triangle ABC$ with $\overline{BD} \perp \overline{AC}$. Describe two geometric means.**

a. $\triangle ADB \sim \triangle BDC$ so $\frac{AD}{BD} = \frac{BD}{CD}$.

In $\triangle ABC$, the altitude is the geometric mean between the two segments of the hypotenuse.

b. $\triangle ABC \sim \triangle ADB$ and $\triangle ABC \sim \triangle BDC$, so $\frac{AC}{AB} = \frac{AB}{AD}$ and $\frac{AC}{BC} = \frac{BC}{DC}$.

In $\triangle ABC$, each leg is the geometric mean between the hypotenuse and the segment of the hypotenuse adjacent to that leg.

Example 2 **Find x, y, and z.**

$\frac{PR}{PQ} = \frac{PQ}{PS}$

$\frac{25}{15} = \frac{15}{x}$ $PR = 25$, $PQ = 15$, $PS = x$

$25x = 225$ Cross multiply.

$x = 9$ Divide each side by 25.

Then

$y = PR - SP$

$= 25 - 9$

$= 16$

$\frac{PR}{QR} = \frac{QR}{RS}$

$\frac{25}{z} = \frac{z}{y}$ $PR = 25$, $QR = z$, $RS = y$

$\frac{25}{z} = \frac{z}{16}$ $y = 16$

$z^2 = 400$ Cross multiply.

$z = 20$ Take the square root of each side.

Exercises

Find x, y, and z to the nearest tenth.

1.

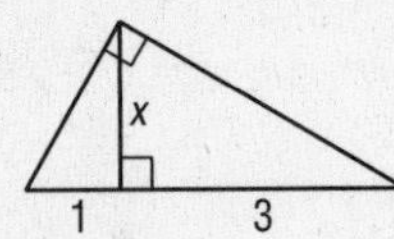

2.

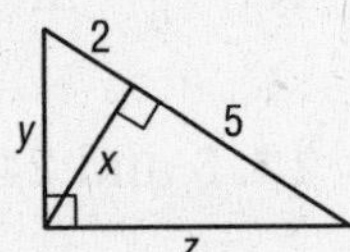

3.

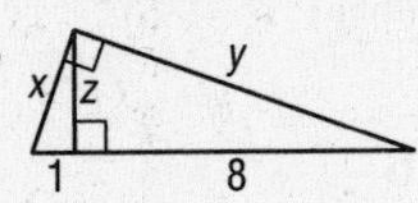

4.

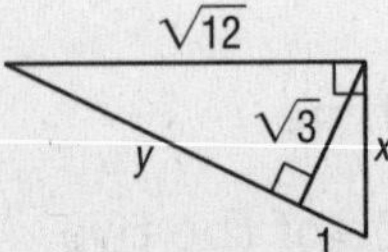

5.

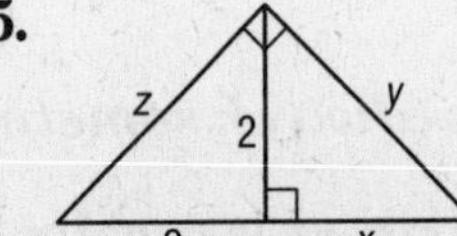

6.

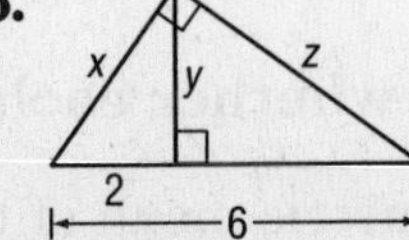

NAME ______________________ DATE ____________ PERIOD _____

7-2 Study Guide and Intervention

The Pythagorean Theorem and Its Converse

The Pythagorean Theorem In a right triangle, the sum of the squares of the measures of the legs equals the square of the measure of the hypotenuse.

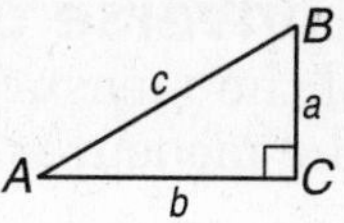

$\triangle ABC$ is a right triangle, so $a^2 + b^2 = c^2$.

Example 1 **Prove the Pythagorean Theorem.**

With altitude $\overline{CD}$, each leg a and b is a geometric mean between hypotenuse c and the segment of the hypotenuse adjacent to that leg.

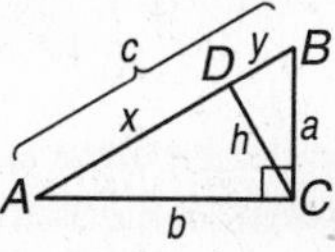

$\frac{c}{a} = \frac{a}{y}$ and $\frac{c}{b} = \frac{b}{x}$, so $a^2 = cy$ and $b^2 = cx$.

Add the two equations and substitute $c = y + x$ to get

$a^2 + b^2 = cy + cx = c(y + x) = c^2$.

Example 2

a. Find a.

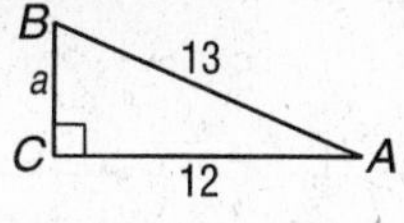

$a^2 + b^2 = c^2$	Pythagorean Theorem
$a^2 + 12^2 = 13^2$	$b = 12, c = 13$
$a^2 + 144 = 169$	Simplify.
$a^2 = 25$	Subtract.
$a = 5$	Take the square root of each side.

b. Find c.

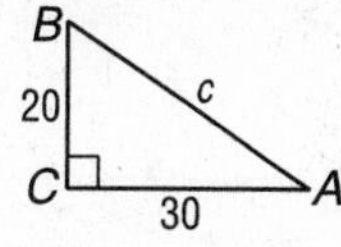

$a^2 + b^2 = c^2$	Pythagorean Theorem
$20^2 + 30^2 = c^2$	$a = 20, b = 30$
$400 + 900 = c^2$	Simplify.
$1300 = c^2$	Add.
$\sqrt{1300} = c$	Take the square root of each side.
$36.1 \approx c$	Use a calculator.

Exercises

Find x.

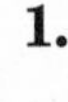
1.

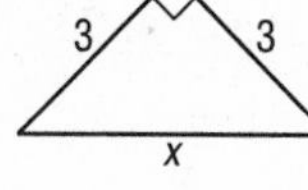

2.

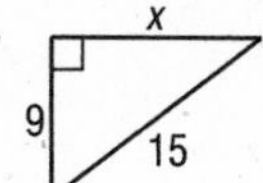

3.

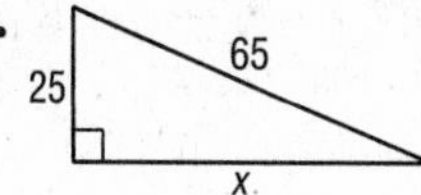

4.

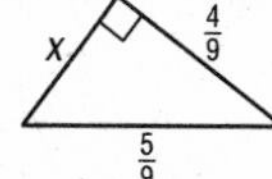

5.

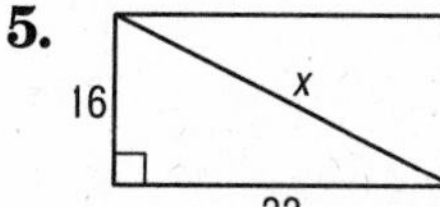

6.

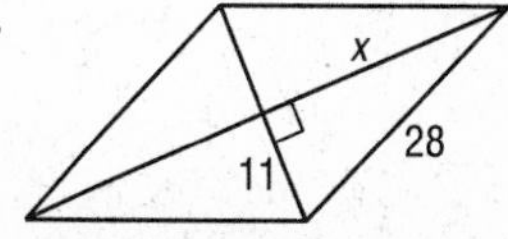

Lesson 7-2

NAME ______________________________ DATE ____________ PERIOD _____

7-2 Study Guide and Intervention *(continued)*

The Pythagorean Theorem and Its Converse

Converse of the Pythagorean Theorem If the sum of the squares of the measures of the two shorter sides of a triangle equals the square of the measure of the longest side, then the triangle is a right triangle.

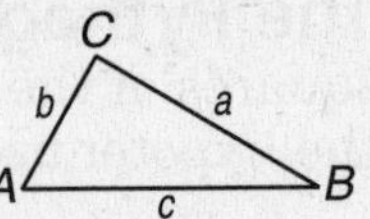

If $a^2 + b^2 = c^2$, then $\triangle ABC$ is a right triangle.

If the three whole numbers a, b, and c satisfy the equation $a^2 + b^2 = c^2$, then the numbers a, b, and c form a **Pythagorean triple**.

Example **Determine whether $\triangle PQR$ is a right triangle.**

$a^2 + b^2 \stackrel{?}{=} c^2$ Pythagorean Theorem

$10^2 + (10\sqrt{3})^2 \stackrel{?}{=} 20^2$ $a = 10$, $b = 10\sqrt{3}$, $c = 20$

$100 + 300 \stackrel{?}{=} 400$ Simplify.

$400 = 400$ ✓ Add.

The sum of the squares of the two shorter sides equals the square of the longest side, so the triangle is a right triangle.

Exercises

Determine whether each set of measures can be the measures of the sides of a right triangle. Then state whether they form a Pythagorean triple.

1. 30, 40, 50

2. 20, 30, 40

3. 18, 24, 30

4. 6, 8, 9

5. $\frac{3}{7}, \frac{4}{7}, \frac{5}{7}$

6. 10, 15, 20

7. $\sqrt{5}, \sqrt{12}, \sqrt{13}$

8. $2, \sqrt{8}, \sqrt{12}$

9. 9, 40, 41

A *family* of Pythagorean triples consists of multiples of known triples. For each Pythagorean triple, find two triples in the same family.

10. 3, 4, 5

11. 5, 12, 13

12. 7, 24, 25

7-3 Study Guide and Intervention

Special Right Triangles

Properties of 45°-45°-90° Triangles The sides of a 45°-45°-90° right triangle have a special relationship.

Example 1 **If the leg of a 45°-45°-90° right triangle is x units, show that the hypotenuse is $x\sqrt{2}$ units.**

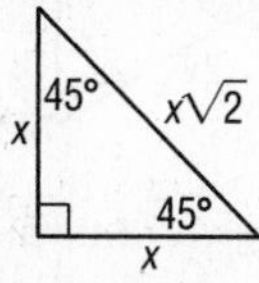

Using the Pythagorean Theorem with $a = b = x$, then

$$\begin{aligned} c^2 &= a^2 + b^2 \\ &= x^2 + x^2 \\ &= 2x^2 \\ c &= \sqrt{2x^2} \\ &= x\sqrt{2} \end{aligned}$$

Example 2 **In a 45°-45°-90° right triangle the hypotenuse is $\sqrt{2}$ times the leg. If the hypotenuse is 6 units, find the length of each leg.**

The hypotenuse is $\sqrt{2}$ times the leg, so divide the length of the hypotenuse by $\sqrt{2}$.

$$\begin{aligned} a &= \frac{6}{\sqrt{2}} \\ &= \frac{6\sqrt{2}}{\sqrt{2}\sqrt{2}} \\ &= \frac{6\sqrt{2}}{2} \\ &= 3\sqrt{2} \text{ units} \end{aligned}$$

Exercises

Find x.

1.

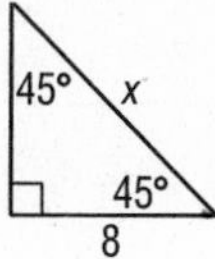

2.

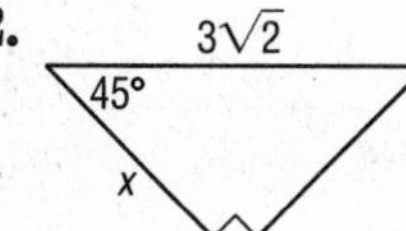

3.

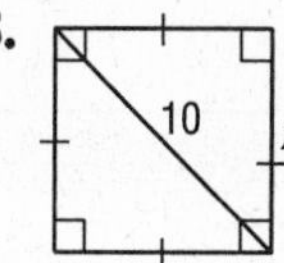

4.

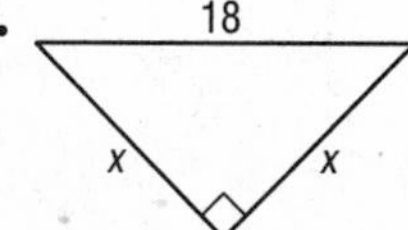

5.

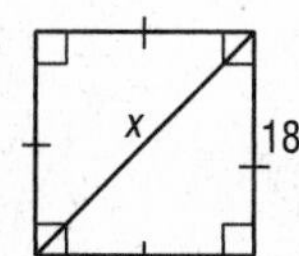

6.

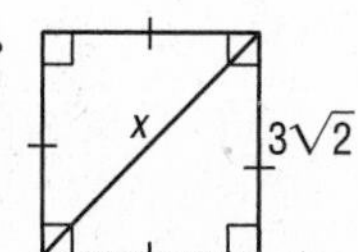

7. Find the perimeter of a square with diagonal 12 centimeters.

8. Find the diagonal of a square with perimeter 20 inches.

9. Find the diagonal of a square with perimeter 28 meters.

Lesson 7-3

7-3 Study Guide and Intervention *(continued)*

Special Right Triangles

Properties of 30°-60°-90° Triangles The sides of a 30°-60°-90° right triangle also have a special relationship.

Example 1 **In a 30°-60°-90° right triangle, show that the hypotenuse is twice the shorter leg and the longer leg is $\sqrt{3}$ times the shorter leg.**

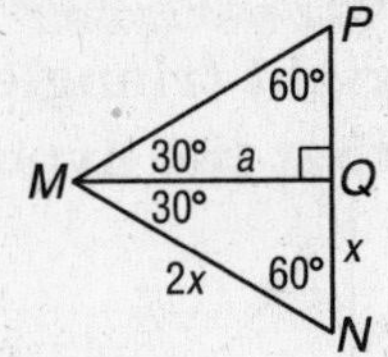

△*MNP* is an equilateral triangle.

△*MNQ* is a 30°-60°-90° right triangle.

△*MNQ* is a 30°-60°-90° right triangle, and the length of the hypotenuse $\overline{MN}$ is two times the length of the shorter side $\overline{NQ}$.
Using the Pythagorean Theorem,

$$\begin{aligned} a^2 &= (2x)^2 - x^2 \\ &= 4x^2 - x^2 \\ &= 3x^2 \\ a &= \sqrt{3x^2} \\ &= x\sqrt{3} \end{aligned}$$

Example 2 **In a 30°-60°-90° right triangle, the hypotenuse is 5 centimeters. Find the lengths of the other two sides of the triangle.**
If the hypotenuse of a 30°-60°-90° right triangle is 5 centimeters, then the length of the shorter leg is half of 5 or 2.5 centimeters. The length of the longer leg is $\sqrt{3}$ times the length of the shorter leg, or $(2.5)(\sqrt{3})$ centimeters.

Exercises

Find *x* and *y*.

1. 60°, $\frac{1}{2}$, x, 30°, y

2. y, 60°, x, 8

3. 11, x, 30°, y

4. x, y, 30°, $9\sqrt{3}$

5.

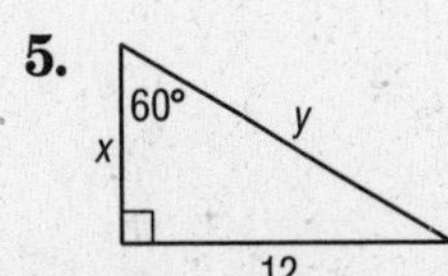

6.

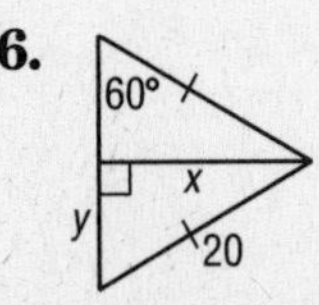

7. The perimeter of an equilateral triangle is 32 centimeters. Find the length of an altitude of the triangle to the nearest tenth of a centimeter.

8. An altitude of an equilateral triangle is 8.3 meters. Find the perimeter of the triangle to the nearest tenth of a meter.

NAME ______________________________ DATE ____________ PERIOD _____

7-4 Study Guide and Intervention

Trigonometry

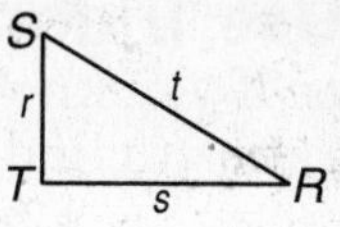

Trigonometric Ratios The ratio of the lengths of two sides of a right triangle is called a **trigonometric ratio**. The three most common ratios are **sine**, **cosine**, and **tangent**, which are abbreviated *sin*, *cos*, and *tan*, respectively.

$$\sin R = \frac{\text{leg opposite } \angle R}{\text{hypotenuse}} = \frac{r}{t} \qquad \cos R = \frac{\text{leg adjacent to } \angle R}{\text{hypotenuse}} = \frac{s}{t} \qquad \tan R = \frac{\text{leg opposite } \angle R}{} = \frac{r}{s}$$

Example **Find sin *A*, cos *A*, and tan *A*. Express each ratio as a decimal to the nearest thousandth.**

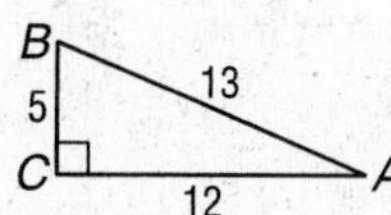

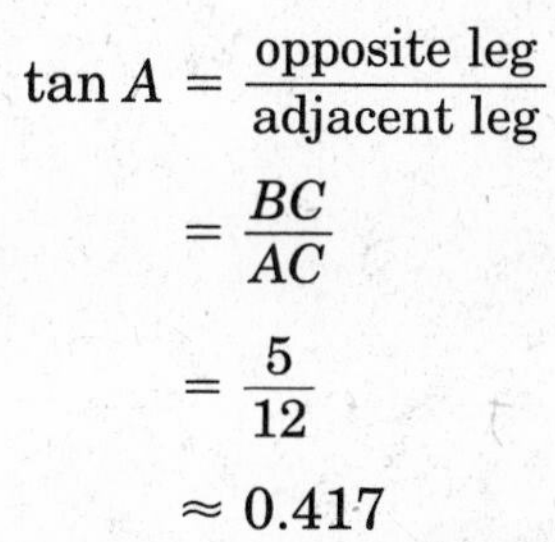

$$\sin A = \frac{\text{opposite leg}}{\text{hypotenuse}} = \frac{BC}{AB} = \frac{5}{13} \approx 0.385$$

$$\cos A = \frac{\text{adjacent leg}}{\text{hypotenuse}} = \frac{AC}{AB} = \frac{12}{13} \approx 0.923$$

$$\tan A = \frac{\text{opposite leg}}{\text{adjacent leg}} = \frac{BC}{AC} = \frac{5}{12} \approx 0.417$$

Exercises

Find the indicated trigonometric ratio as a fraction and as a decimal. If necessary, round to the nearest ten-thousandth.

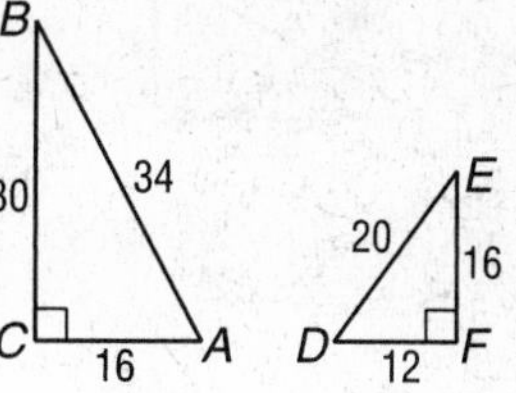

1. sin *A*

2. tan *B*

3. cos *A*

4. cos *B*

5. sin *D*

6. tan *E*

7. cos *E*

8. cos *D*

Lesson 7-4

© Glencoe/McGraw-Hill

7-4 Study Guide and Intervention *(continued)*

Trigonometry

Use Trigonometric Ratios In a right triangle, if you know the measures of two sides or if you know the measures of one side and an acute angle, then you can use trigonometric ratios to find the measures of the missing sides or angles of the triangle.

Example **Find x, y, and z. Round each measure to the nearest whole number.**

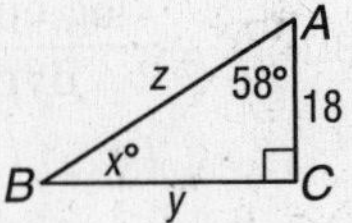

a. Find x.

$$x + 58 = 90$$
$$x = 32$$

b. Find y.

$$\tan A = \frac{y}{18}$$
$$\tan 58° = \frac{y}{18}$$
$$y = 18 \tan 58°$$
$$y \approx 29$$

c. Find z.

$$\cos A = \frac{18}{z}$$
$$\cos 58° = \frac{18}{z}$$
$$z \cos 58° = 18$$
$$z = \frac{18}{\cos 58°}$$
$$z \approx 34$$

Exercises

Find x. Round to the nearest tenth.

1.

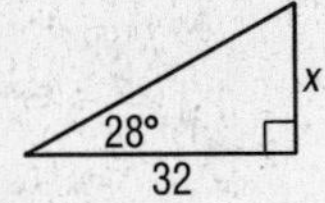

2.

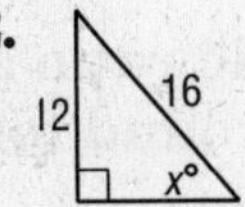

3.

12
x°
5

4.

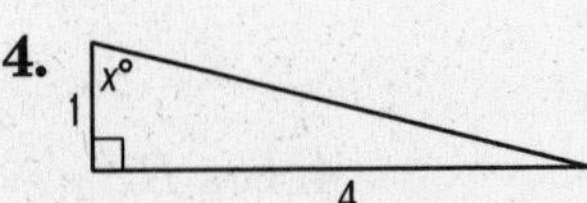

5.

40°
16
x

6.

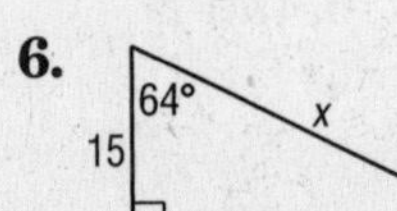

7-5 Study Guide and Intervention

Angles of Elevation and Depression

Angles of Elevation Many real-world problems that involve looking up to an object can be described in terms of an **angle of elevation**, which is the angle between an observer's line of sight and a horizontal line.

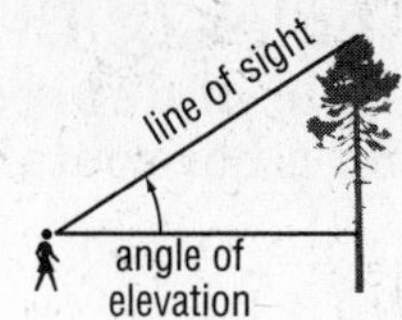

Example **The angle of elevation from point A to the top of a cliff is 34°. If point A is 1000 feet from the base of the cliff, how high is the cliff?**

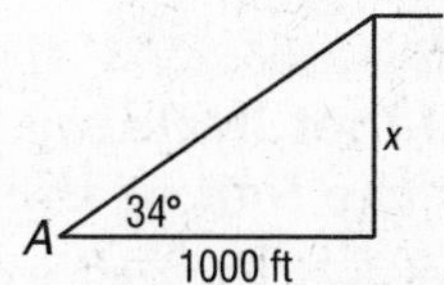

Let x = the height of the cliff.

$\tan 34° = \frac{x}{1000}$ $\quad$ $\tan = \frac{\text{opposite}}{\text{adjacent}}$

$1000(\tan 34°) = x$ $\quad$ Multiply each side by 1000.

$674.5 = x$ $\quad$ Use a calculator.

The height of the cliff is about 674.5 feet.

Exercises

Solve each problem. Round measures of segments to the nearest whole number and angles to the nearest degree.

1. The angle of elevation from point A to the top of a hill is 49°. If point A is 400 feet from the base of the hill, how high is the hill?

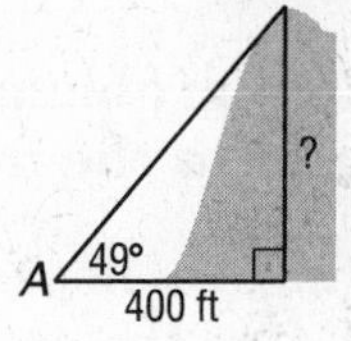

2. Find the angle of elevation of the sun when a 12.5-meter-tall telephone pole casts an 18-meter-long shadow.

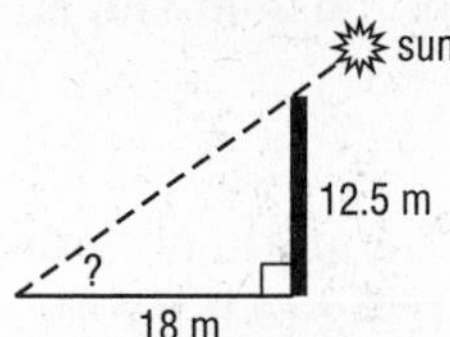

3. A ladder leaning against a building makes an angle of 78° with the ground. The foot of the ladder is 5 feet from the building. How long is the ladder?

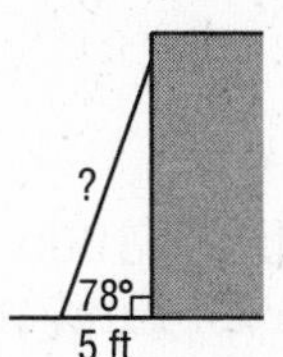

4. A person whose eyes are 5 feet above the ground is standing on the runway of an airport 100 feet from the control tower. That person observes an air traffic controller at the window of the 132-foot tower. What is the angle of elevation?

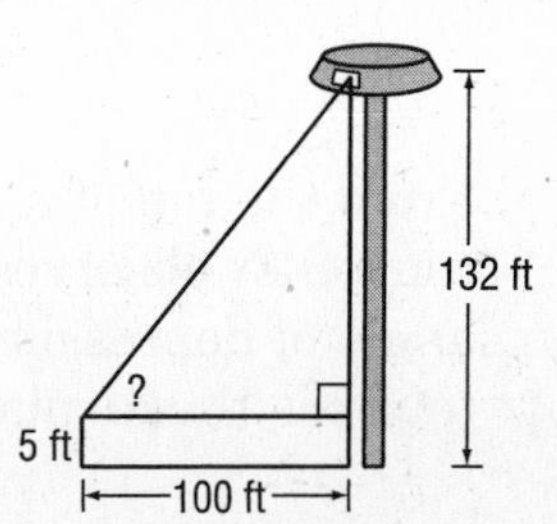

Lesson 7-5

NAME ______________________ DATE ____________ PERIOD ______

7-5 Study Guide and Intervention *(continued)*

Angles of Elevation and Depression

Angles of Depression When an observer is looking down, the **angle of depression** is the angle between the observer's line of sight and a horizontal line.

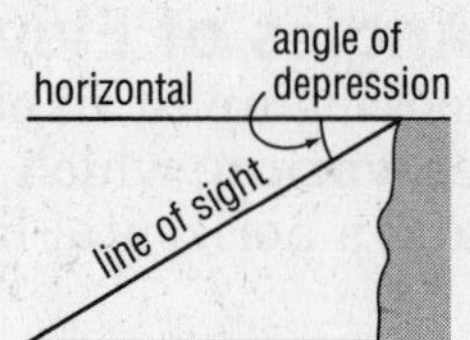

Example **The angle of depression from the top of an 80-foot building to point A on the ground is 42°. How far is the foot of the building from point A?**

horizontal D, B, angle of depression, 80 ft, A, 42°, C, x

Let x = the distance from point A to the foot of the building. Since the horizontal line is parallel to the ground, the angle of depression $\angle DBA$ is congruent to $\angle BAC$.

$\tan 42° = \frac{80}{x}$	$\tan = \frac{\text{opposite}}{\text{adjacent}}$
$x(\tan 42°) = 80$	Multiply each side by x.
$x = \frac{80}{\tan 42°}$	Divide each side by tan 42°.
$x \approx 88.8$	Use a calculator.

Point A is about 89 feet from the base of the building.

Exercises

Solve each problem. Round measures of segments to the nearest whole number and angles to the nearest degree.

1. The angle of depression from the top of a sheer cliff to point A on the ground is 35°. If point A is 280 feet from the base of the cliff, how tall is the cliff?

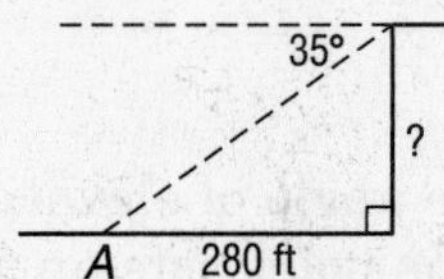

2. The angle of depression from a balloon on a 75-foot string to a person on the ground is 36°. How high is the balloon?

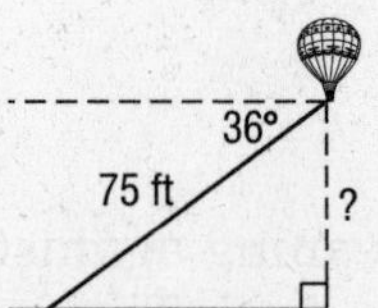

3. A ski run is 1000 yards long with a vertical drop of 208 yards. Find the angle of depression from the top of the ski run to the bottom.

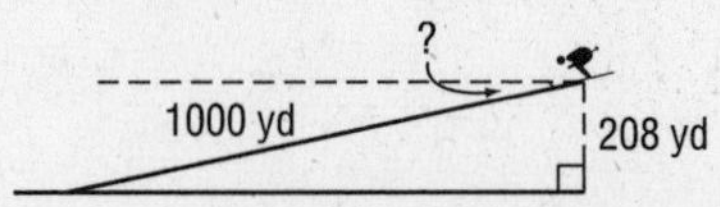

4. From the top of a 120-foot-high tower, an air traffic controller observes an airplane on the runway at an angle of depression of 19°. How far from the base of the tower is the airplane?

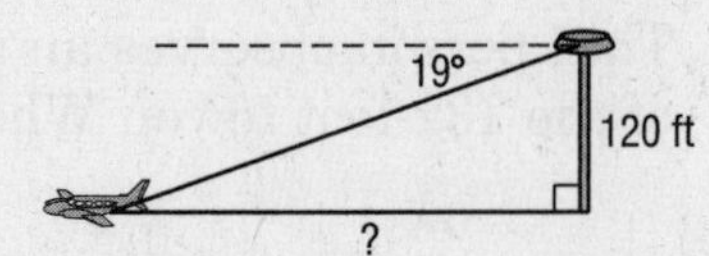

© Glencoe/McGraw-Hill

7-6 Study Guide and Intervention

The Law of Sines

The Law of Sines In any triangle, there is a special relationship between the angles of the triangle and the lengths of the sides opposite the angles.

Law of Sines	$\frac{\sin A}{a} = \frac{\sin B}{b} = \frac{\sin C}{c}$

Example 1 **In $\triangle ABC$, find b.**

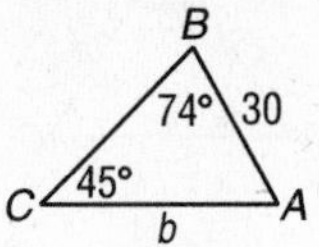

$\frac{\sin C}{c} = \frac{\sin B}{b}$ Law of Sines

$\frac{\sin 45°}{30} = \frac{\sin 74°}{b}$ $m\angle C = 45$, $c = 30$, $m\angle B = 74$

$b \sin 45° = 30 \sin 74°$ Cross multiply.

$b = \frac{30 \sin 74°}{\sin 45°}$ Divide each side by sin 45°.

$b \approx 40.8$ Use a calculator.

Example 2 **In $\triangle DEF$, find $m\angle D$.**

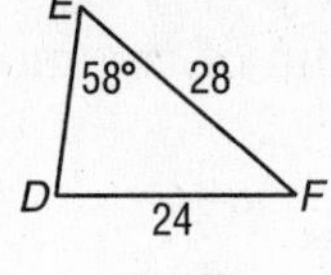

$\frac{\sin D}{d} = \frac{\sin E}{e}$ Law of Sines

$\frac{\sin D}{28} = \frac{\sin 58°}{24}$ $d = 28$, $m\angle E = 58$, $e = 24$

$24 \sin D = 28 \sin 58°$ Cross multiply.

$\sin D = \frac{28 \sin 58°}{24}$ Divide each side by 24.

$D = \sin^{-1} \frac{28 \sin 58°}{24}$ Use the inverse sine.

$D \approx 81.6°$ Use a calculator.

Exercises

Find each measure using the given measures of $\triangle ABC$. Round angle measures to the nearest degree and side measures to the nearest tenth.

1. If $c = 12$, $m\angle A = 80$, and $m\angle C = 40$, find a.

2. If $b = 20$, $c = 26$, and $m\angle C = 52$, find $m\angle B$.

3. If $a = 18$, $c = 16$, and $m\angle A = 84$, find $m\angle C$.

4. If $a = 25$, $m\angle A = 72$, and $m\angle B = 17$, find b.

5. If $b = 12$, $m\angle A = 89$, and $m\angle B = 80$, find a.

6. If $a = 30$, $c = 20$, and $m\angle A = 60$, find $m\angle C$.

7-6 Study Guide and Intervention *(continued)*

The Law of Sines

Use the Law of Sines to Solve Problems You can use the **Law of Sines** to solve some problems that involve triangles.

Law of Sines	Let $\triangle ABC$ be any triangle with *a*, *b*, and *c* representing the measures of the sides opposite the angles with measures *A*, *B*, and *C*, respectively. Then $\frac{\sin A}{a} = \frac{\sin B}{b} = \frac{\sin C}{c}$.

Example **Isosceles $\triangle ABC$ has a base of 24 centimeters and a vertex angle of 68°. Find the perimeter of the triangle.**

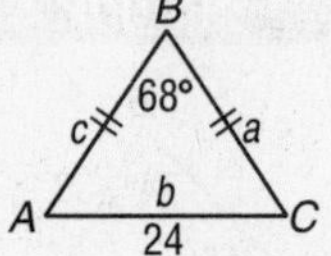

The vertex angle is 68°, so the sum of the measures of the base angles is 112 and $m\angle A = m\angle C = 56$.

$\frac{\sin B}{b} = \frac{\sin A}{a}$ Law of Sines

$\frac{\sin 68°}{24} = \frac{\sin 56°}{a}$ $m\angle B = 68$, $b = 24$, $m\angle A = 56$

$a \sin 68° = 24 \sin 56°$ Cross multiply.

$a = \frac{24 \sin 56°}{\sin 68°}$ Divide each side by sin 68°.

≈ 21.5 Use a calculator.

The triangle is isosceles, so $c = 21.5$.
The perimeter is 24 + 21.5 + 21.5 or about 67 centimeters.

Exercises

Draw a triangle to go with each exercise and mark it with the given information. Then solve the problem. Round angle measures to the nearest degree and side measures to the nearest tenth.

1. One side of a triangular garden is 42.0 feet. The angles on each end of this side measure 66° and 82°. Find the length of fence needed to enclose the garden.

2. Two radar stations *A* and *B* are 32 miles apart. They locate an airplane *X* at the same time. The three points form $\angle XAB$, which measures 46°, and $\angle XBA$, which measures 52°. How far is the airplane from each station?

3. A civil engineer wants to determine the distances from points *A* and *B* to an inaccessible point *C* in a river. $\angle BAC$ measures 67° and $\angle ABC$ measures 52°. If points *A* and *B* are 82.0 feet apart, find the distance from *C* to each point.

4. A ranger tower at point *A* is 42 kilometers north of a ranger tower at point *B*. A fire at point *C* is observed from both towers. If $\angle BAC$ measures 43° and $\angle ABC$ measures 68°, which ranger tower is closer to the fire? How much closer?

© Glencoe/McGraw-Hill

7-7 Study Guide and Intervention

The Law of Cosines

The Law of Cosines Another relationship between the sides and angles of any triangle is called the **Law of Cosines**. You can use the Law of Cosines if you know three sides of a triangle or if you know two sides and the included angle of a triangle.

Law of Cosines	Let $\triangle ABC$ be any triangle with a, b, and c representing the measures of the sides opposite the angles with measures A, B, and C, respectively. Then the following equations are true. $a^2 = b^2 + c^2 - 2bc \cos A$ $\quad b^2 = a^2 + c^2 - 2ac \cos B$ $\quad c^2 = a^2 + b^2 - 2ab \cos C$

Example 1 **In $\triangle ABC$, find c.**

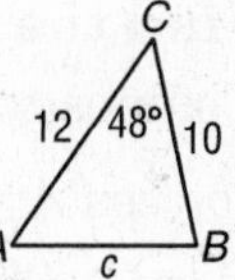

$c^2 = a^2 + b^2 - 2ab \cos C$ — Law of Cosines

$c^2 = 12^2 + 10^2 - 2(12)(10)\cos 48°$ — $a = 12$, $b = 10$, $m\angle C = 48$

$c = \sqrt{12^2 + 10^2 - 2(12)(10)\cos 48°}$ — Take the square root of each side.

$c \approx 9.1$ — Use a calculator.

Example 2 **In $\triangle ABC$, find $m\angle A$.**

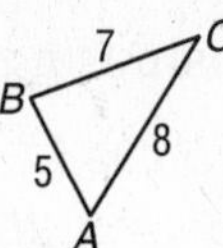

$a^2 = b^2 + c^2 - 2bc \cos A$ — Law of Cosines

$7^2 = 5^2 + 8^2 - 2(5)(8) \cos A$ — $a = 7$, $b = 5$, $c = 8$

$49 = 25 + 64 - 80 \cos A$ — Multiply.

$-40 = -80 \cos A$ — Subtract 89 from each side.

$\frac{1}{2} = \cos A$ — Divide each side by -80.

$\cos^{-1}\frac{1}{2} = A$ — Use the inverse cosine.

$60° = A$ — Use a calculator.

Exercises

Find each measure using the given measures from $\triangle ABC$. Round angle measures to the nearest degree and side measures to the nearest tenth.

1. If $b = 14$, $c = 12$, and $m\angle A = 62$, find a.

2. If $a = 11$, $b = 10$, and $c = 12$, find $m\angle B$.

3. If $a = 24$, $b = 18$, and $c = 16$, find $m\angle C$.

4. If $a = 20$, $c = 25$, and $m\angle B = 82$, find b.

5. If $b = 18$, $c = 28$, and $m\angle A = 59$, find a.

6. If $a = 15$, $b = 19$, and $c = 15$, find $m\angle C$.

7-7 Study Guide and Intervention *(continued)*

The Law of Cosines

Use the Law of Cosines to Solve Problems

You can use the **Law of Cosines** to solve some problems involving triangles.

Law of Cosines	Let $\triangle ABC$ be any triangle with *a*, *b*, and *c* representing the measures of the sides opposite the angles with measures *A*, *B*, and *C*, respectively. Then the following equations are true. $a^2 = b^2 + c^2 - 2bc \cos A$ $b^2 = a^2 + c^2 - 2ac \cos B$ $c^2 = a^2 + b^2 - 2ab \cos C$

Example **Ms. Jones wants to purchase a piece of land with the shape shown. Find the perimeter of the property.**

300 ft
80°
300 ft
a
c
88°
200 ft

Use the Law of Cosines to find the value of a.

$a^2 = b^2 + c^2 - 2bc \cos A$ — Law of Cosines

$a^2 = 300^2 + 200^2 - 2(300)(200) \cos 88°$ — $b = 300$, $c = 200$, $m\angle A = 88$

$a = \sqrt{130{,}000 - 120{,}000 \cos 88°}$ — Take the square root of each side.

≈ 354.7 — Use a calculator.

Use the Law of Cosines again to find the value of c.

$c^2 = a^2 + b^2 - 2ab \cos C$ — Law of Cosines

$c^2 = 354.7^2 + 300^2 - 2(354.7)(300) \cos 80°$ — $a = 354.7$, $b = 300$, $m\angle C = 80$

$c = \sqrt{215{,}812.09 - 212{,}820 \cos 80°}$ — Take the square root of each side.

≈ 422.9 — Use a calculator.

The perimeter of the land is 300 + 200 + 422.9 + 200 or about 1223 feet.

Exercises

Draw a figure or diagram to go with each exercise and mark it with the given information. Then solve the problem. Round angle measures to the nearest degree and side measures to the nearest tenth.

1. A triangular garden has dimensions 54 feet, 48 feet, and 62 feet. Find the angles at each corner of the garden.

2. A parallelogram has a 68° angle and sides 8 and 12. Find the lengths of the diagonals.

3. An airplane is sighted from two locations, and its position forms an acute triangle with them. The distance to the airplane is 20 miles from one location with an angle of elevation 48.0°, and 40 miles from the other location with an angle of elevation of 21.8°. How far apart are the two locations?

4. A ranger tower at point A is directly north of a ranger tower at point B. A fire at point C is observed from both towers. The distance from the fire to tower A is 60 miles, and the distance from the fire to tower B is 50 miles. If $m\angle ACB = 62$, find the distance between the towers.

NAME ______________________ DATE ____________ PERIOD _____

8-1 Study Guide and Intervention

Angles of Polygons

Sum of Measures of Interior Angles The segments that connect the nonconsecutive sides of a polygon are called **diagonals**. Drawing all of the diagonals from one vertex of an ***n*-gon** separates the polygon into $n - 2$ triangles. The sum of the measures of the interior angles of the polygon can be found by adding the measures of the interior angles of those $n - 2$ triangles.

Interior Angle Sum Theorem	If a convex polygon has n sides, and S is the sum of the measures of its interior angles, then $S = 180(n - 2)$.

Lesson 8-1

Example 1 **A convex polygon has 13 sides. Find the sum of the measures of the interior angles.**

$S = 180(n - 2)$
$= 180(13 - 2)$
$= 180(11)$
$= 1980$

Example 2 **The measure of an interior angle of a regular polygon is 120. Find the number of sides.**

The number of sides is n, so the sum of the measures of the interior angles is $120n$.

$S = 180(n - 2)$
$120n = 180(n - 2)$
$120n = 180n - 360$
$-60n = -360$
$n = 6$

Exercises

Find the sum of the measures of the interior angles of each convex polygon.

1. 10-gon **2.** 16-gon **3.** 30-gon

4. 8-gon **5.** 12-gon **6.** $3x$-gon

The measure of an interior angle of a regular polygon is given. Find the number of sides in each polygon.

7. 150 **8.** 160 **9.** 175

10. 165 **11.** 168.75 **12.** 135

13. Find x.

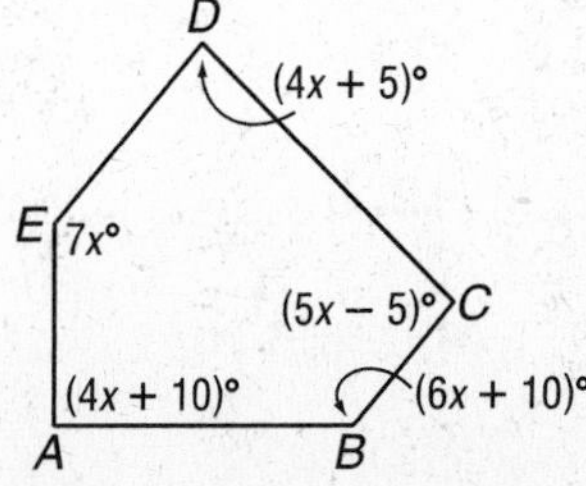

8-1 Study Guide and Intervention *(continued)*

Angles of Polygons

Sum of Measures of Exterior Angles There is a simple relationship among the exterior angles of a convex polygon.

Exterior Angle Sum Theorem	If a polygon is convex, then the sum of the measures of the exterior angles, one at each vertex, is 360.

Example 1 **Find the sum of the measures of the exterior angles, one at each vertex, of a convex 27-gon.**

For *any* convex polygon, the sum of the measures of its exterior angles, one at each vertex, is 360.

Example 2 **Find the measure of each exterior angle of regular hexagon *ABCDEF*.**

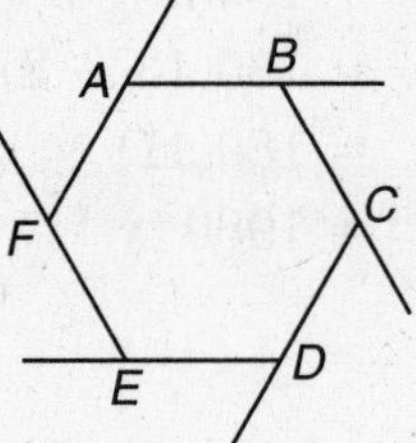

The sum of the measures of the exterior angles is 360 and a hexagon has 6 angles. If n is the measure of each exterior angle, then

$6n = 360$

$n = 60$

Exercises

Find the sum of the measures of the exterior angles of each convex polygon.

1. 10-gon **2.** 16-gon **3.** 36-gon

Find the measure of an exterior angle for each convex regular polygon.

4. 12-gon **5.** 36-gon **6.** $2x$-gon

Find the measure of an exterior angle given the number of sides of a regular polygon.

7. 40 **8.** 18 **9.** 12

10. 24 **11.** 180 **12.** 8

NAME ______________________ DATE __________ PERIOD _____

8-2 Study Guide and Intervention

Parallelograms

Sides and Angles of Parallelograms A quadrilateral with both pairs of opposite sides parallel is a **parallelogram**. Here are four important properties of parallelograms.

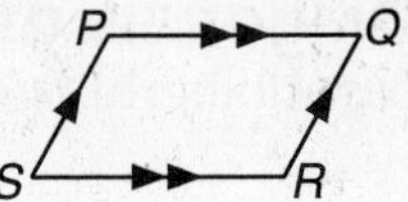

	If *PQRS* is a parallelogram, then
The opposite sides of a parallelogram are congruent.	$\overline{PQ} \cong \overline{SR}$ and $\overline{PS} \cong \overline{QR}$
The opposite angles of a parallelogram are congruent.	$\angle P \cong \angle R$ and $\angle S \cong \angle Q$
The consecutive angles of a parallelogram are supplementary.	$\angle P$ and $\angle S$ are supplementary; $\angle S$ and $\angle R$ are supplementary; $\angle R$ and $\angle Q$ are supplementary; $\angle Q$ and $\angle P$ are supplementary.
If a parallelogram has one right angle, then it has four right angles.	If $m\angle P = 90$, then $m\angle Q = 90$, $m\angle R = 90$, and $m\angle S = 90$.

Example **If *ABCD* is a parallelogram, find *a* and *b*.**

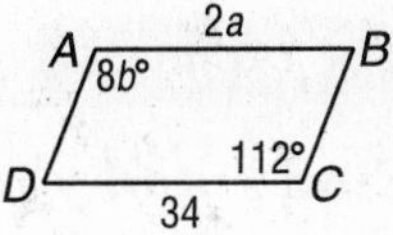

$\overline{AB}$ and $\overline{CD}$ are opposite sides, so $\overline{AB} \cong \overline{CD}$.

$2a = 34$

$a = 17$

$\angle A$ and $\angle C$ are opposite angles, so $\angle A \cong \angle C$.

$8b = 112$

$b = 14$

Exercises

Find *x* and *y* in each parallelogram.

1.

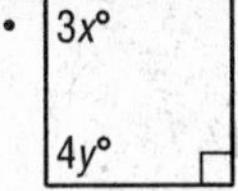

2.

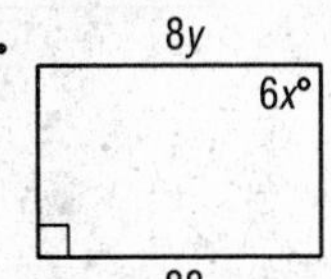

3.

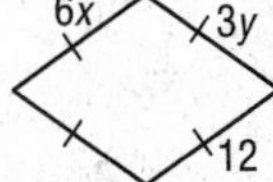

4.

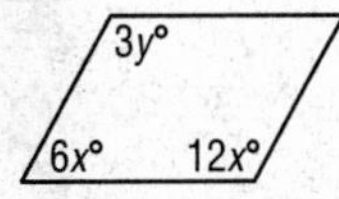

5.

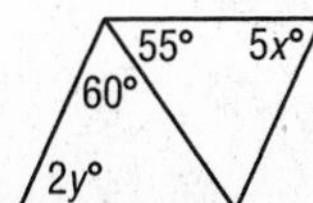

6. 2y; 30x; 150; 72x

8-2 Study Guide and Intervention *(continued)*

Parallelograms

Diagonals of Parallelograms Two important properties of parallelograms deal with their diagonals.

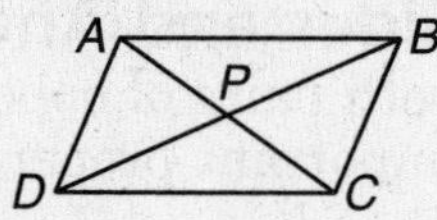

	If *ABCD* is a parallelogram, then:
The diagonals of a parallelogram bisect each other.	$AP = PC$ and $DP = PB$
Each diagonal separates a parallelogram into two congruent triangles.	$\triangle ACD \cong \triangle CAB$ and $\triangle ADB \cong \triangle CBD$

Example **Find x and y in parallelogram *ABCD*.**

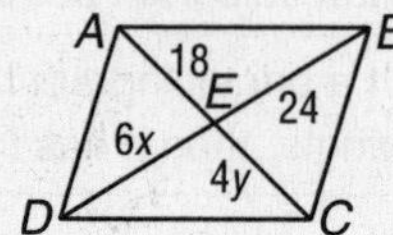

The diagonals bisect each other, so $AE = CE$ and $DE = BE$.

$6x = 24$ $\quad$ $4y = 18$

$x = 4$ $\quad$ $y = 4.5$

Exercises

Find x and y in each parallelogram.

1.

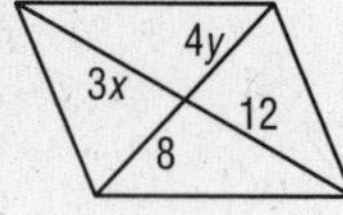

2.

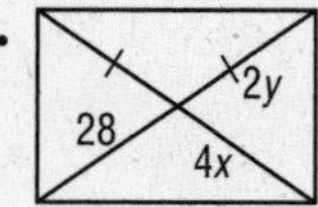

3.

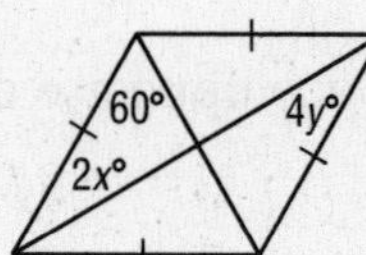

4.

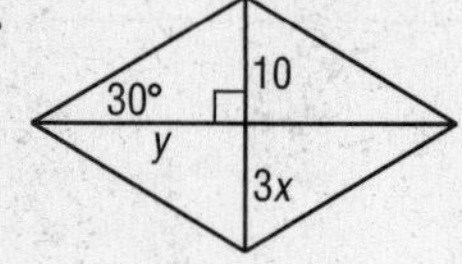

5.

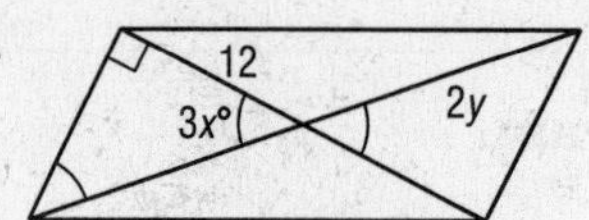

6.

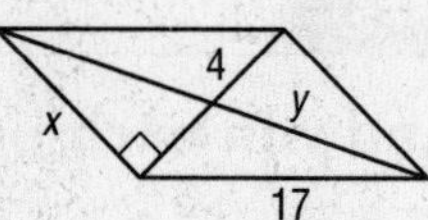

Complete each statement about ▱*ABCD*. Justify your answer.

7. $\angle BAC \cong$

8. $\overline{DE} \cong$

9. $\triangle ADC \cong$

10. $\overline{AD} \parallel$

NAME ______________________ DATE __________ PERIOD _____

8-3 Study Guide and Intervention

Tests for Parallelograms

Conditions for a Parallelogram There are many ways to establish that a quadrilateral is a parallelogram.

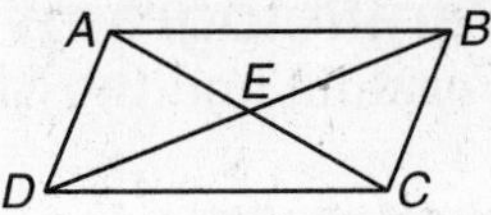

If:	If:
both pairs of opposite sides are parallel,	$\overline{AB} \parallel \overline{DC}$ and $\overline{AD} \parallel \overline{BC}$,
both pairs of opposite sides are congruent,	$\overline{AB} \cong \overline{DC}$ and $\overline{AD} \cong \overline{BC}$,
both pairs of opposite angles are congruent,	$\angle ABC \cong \angle ADC$ and $\angle DAB \cong \angle BCD$,
the diagonals bisect each other,	$\overline{AE} \cong \overline{CE}$ and $\overline{DE} \cong \overline{BE}$,
one pair of opposite sides is congruent and parallel,	$\overline{AB} \parallel \overline{CD}$ and $\overline{AB} \cong \overline{CD}$, or $\overline{AD} \parallel \overline{BC}$ and $\overline{AD} \cong \overline{BC}$,
then: the figure is a parallelogram.	**then:** $ABCD$ is a parallelogram.

Example **Find x and y so that $FGHJ$ is a parallelogram.**

$FGHJ$ is a parallelogram if the lengths of the opposite sides are equal.

F, G, H, J; $6x + 3$; $4x - 2y$; 2; 15

$$6x + 3 = 15 \qquad 4x - 2y = 2$$
$$6x = 12 \qquad 4(2) - 2y = 2$$
$$x = 2 \qquad 8 - 2y = 2$$
$$-2y = -6$$
$$y = 3$$

Exercises

Find x and y so that each quadrilateral is a parallelogram.

1.

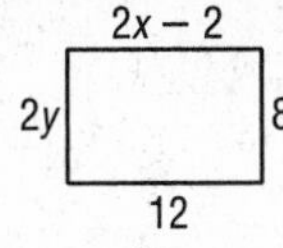

2.

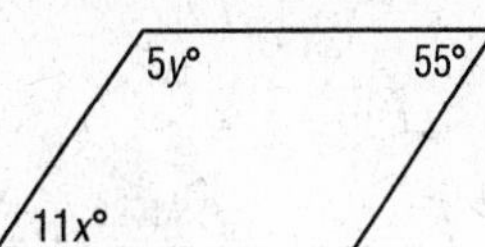

3.

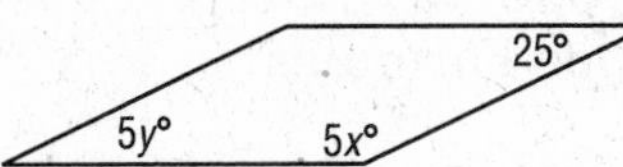

4.

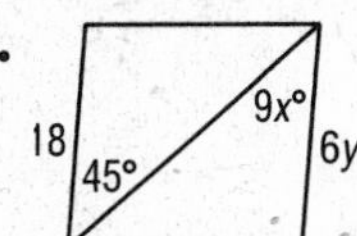

5.

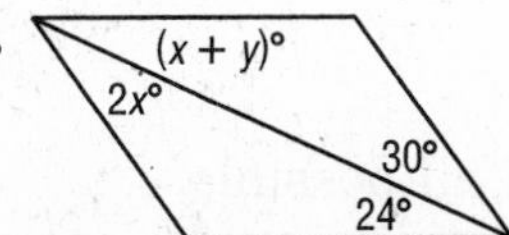

6.

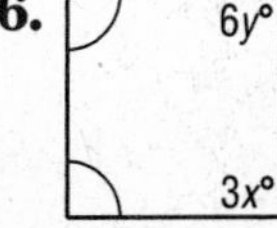

Lesson 8-3

8-3 Study Guide and Intervention *(continued)*

Tests for Parallelograms

Parallelograms on the Coordinate Plane On the coordinate plane, the Distance Formula and the Slope Formula can be used to test if a quadrilateral is a parallelogram.

Example **Determine whether *ABCD* is a parallelogram.**

The vertices are $A(-2, 3)$, $B(3, 2)$, $C(2, -1)$, and $D(-3, 0)$.

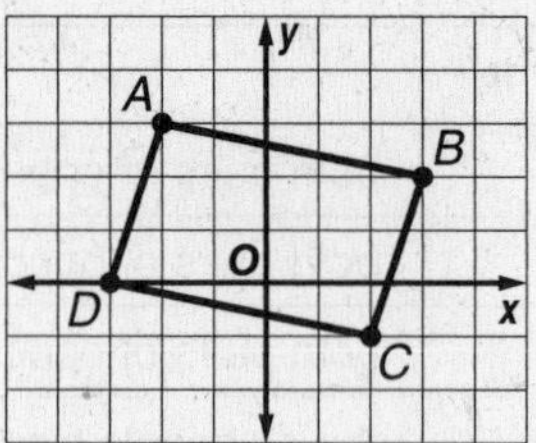

Method 1: Use the Slope Formula, $m = \frac{y_2 - y_1}{x_2 - x_1}$.

slope of $\overline{AD} = \frac{3 - 0}{-2 - (-3)} = \frac{3}{1} = 3$ slope of $\overline{BC} = \frac{2 - (-1)}{3 - 2} = \frac{3}{1} = 3$

slope of $\overline{AB} = \frac{2 - 3}{3 - (-2)} = -\frac{1}{5}$ slope of $\overline{CD} = \frac{-1 - 0}{2 - (-3)} = -\frac{1}{5}$

Opposite sides have the same slope, so $\overline{AB} \parallel \overline{CD}$ and $\overline{AD} \parallel \overline{BC}$. Both pairs of opposite sides are parallel, so *ABCD* is a parallelogram.

Method 2: Use the Distance Formula, $d = \sqrt{(x_2 - x_1)^2 + (y_2 - y_1)^2}$.

$AB = \sqrt{(-2 - 3)^2 + (3 - 2)^2} = \sqrt{25 + 1}$ or $\sqrt{26}$

$CD = \sqrt{(2 - (-3))^2 + (-1 - 0)^2} = \sqrt{25 + 1}$ or $\sqrt{26}$

$AD = \sqrt{(-2 - (-3))^2 + (3 - 0)^2} = \sqrt{1 + 9}$ or $\sqrt{10}$

$BC = \sqrt{(3 - 2)^2 + (2 - (-1))^2} = \sqrt{1 + 9}$ or $\sqrt{10}$

Both pairs of opposite sides have the same length, so *ABCD* is a parallelogram.

Exercises

Determine whether a figure with the given vertices is a parallelogram. Use the method indicated.

1. $A(0, 0)$, $B(1, 3)$, $C(5, 3)$, $D(4, 0)$;
Slope Formula

2. $D(-1, 1)$, $E(2, 4)$, $F(6, 4)$, $G(3, 1)$;
Slope Formula

3. $R(-1, 0)$, $S(3, 0)$, $T(2, -3)$, $U(-3, -2)$;
Distance Formula

4. $A(-3, 2)$, $B(-1, 4)$, $C(2, 1)$, $D(0, -1)$;
Distance and Slope Formulas

5. $S(-2, 4)$, $T(-1, -1)$, $U(3, -4)$, $V(2, 1)$;
Distance and Slope Formulas

6. $F(3, 3)$, $G(1, 2)$, $H(-3, 1)$, $I(-1, 4)$;
Midpoint Formula

7. A parallelogram has vertices $R(-2, -1)$, $S(2, 1)$, and $T(0, -3)$. Find all possible coordinates for the fourth vertex.

NAME ______________________ DATE __________ PERIOD _____

8-4 Study Guide and Intervention

Rectangles

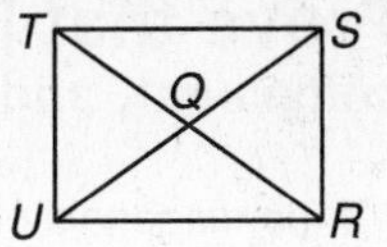

Properties of Rectangles A **rectangle** is a quadrilateral with four right angles. Here are the properties of rectangles.

A rectangle has all the properties of a parallelogram.

- Opposite sides are parallel.
- Opposite angles are congruent.
- Opposite sides are congruent.
- Consecutive angles are supplementary.
- The diagonals bisect each other.

Also:

- All four angles are right angles. $\angle UTS$, $\angle TSR$, $\angle SRU$, and $\angle RUT$ are right angles.
- The diagonals are congruent. $\overline{TR} \cong \overline{US}$

Example 1 **In rectangle *RSTU* above, $US = 6x + 3$ and $RT = 7x - 2$. Find x.**

The diagonals of a rectangle bisect each other, so $US = RT$.

$$6x + 3 = 7x - 2$$
$$3 = x - 2$$
$$5 = x$$

Example 2 **In rectangle *RSTU* above, $m\angle STR = 8x + 3$ and $m\angle UTR = 16x - 9$. Find $m\angle STR$.**

$\angle UTS$ is a right angle, so $m\angle STR + m\angle UTR = 90$.

$$8x + 3 + 16x - 9 = 90$$
$$24x - 6 = 90$$
$$24x = 96$$
$$x = 4$$

$m\angle STR = 8x + 3 = 8(4) + 3$ or 35

Exercises

***ABCD* is a rectangle.**

1. If $AE = 36$ and $CE = 2x - 4$, find x.
2. If $BE = 6y + 2$ and $CE = 4y + 6$, find y.
3. If $BC = 24$ and $AD = 5y - 1$, find y.
4. If $m\angle BEA = 62$, find $m\angle BAC$.
5. If $m\angle AED = 12x$ and $m\angle BEC = 10x + 20$, find $m\angle AED$.
6. If $BD = 8y - 4$ and $AC = 7y + 3$, find BD.
7. If $m\angle DBC = 10x$ and $m\angle ACB = 4x^2 - 6$, find $m\angle ACB$.
8. If $AB = 6y$ and $BC = 8y$, find BD in terms of y.

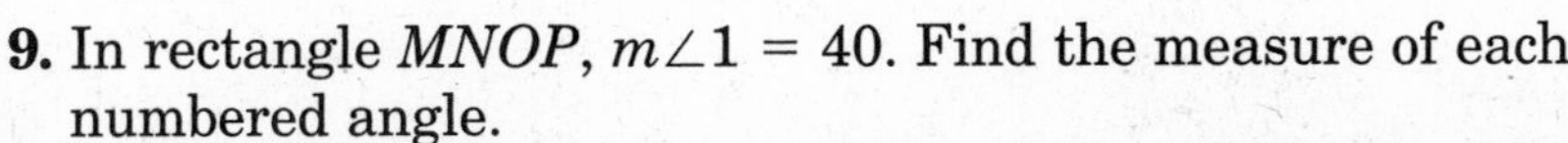

9. In rectangle *MNOP*, $m\angle 1 = 40$. Find the measure of each numbered angle.

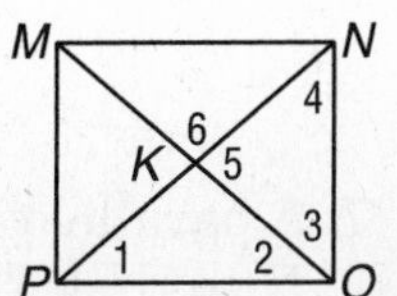

Lesson 8-4

8-4 Study Guide and Intervention *(continued)*

Rectangles

Prove that Parallelograms Are Rectangles The diagonals of a rectangle are congruent, and the converse is also true.

If the diagonals of a parallelogram are congruent, then the parallelogram is a rectangle.

In the coordinate plane you can use the Distance Formula, the Slope Formula, and properties of diagonals to show that a figure is a rectangle.

Example **Determine whether $A(-3, 0)$, $B(-2, 3)$, $C(4, 1)$, and $D(3, -2)$ are the vertices of a rectangle.**

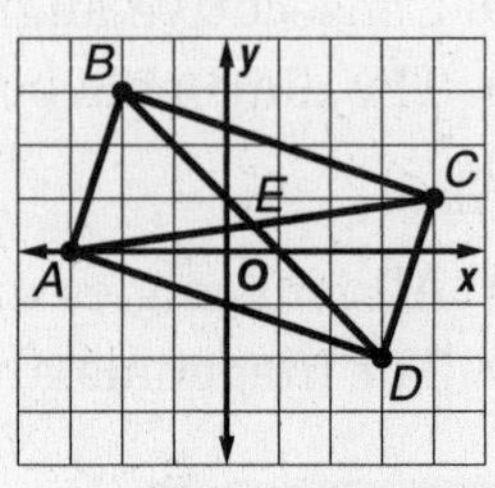

Method 1: Use the Slope Formula.

slope of $\overline{AB} = \frac{3 - 0}{-2 - (-3)} = \frac{3}{1}$ or 3 slope of $\overline{AD} = \frac{-2 - 0}{3 - (-3)} = \frac{-2}{6}$ or $-\frac{1}{3}$

slope of $\overline{CD} = \frac{-2 - 1}{3 - 4} = \frac{-3}{-1}$ or 3 slope of $\overline{BC} = \frac{1 - 3}{4 - (-2)} = \frac{-2}{6}$ or $-\frac{1}{3}$

Opposite sides are parallel, so the figure is a parallelogram. Consecutive sides are perpendicular, so *ABCD* is a rectangle.

Method 2: Use the Midpoint and Distance Formulas.

The midpoint of $\overline{AC}$ is $\left(\frac{-3 + 4}{2}, \frac{0 + 1}{2}\right) = \left(\frac{1}{2}, \frac{1}{2}\right)$ and the midpoint of $\overline{BD}$ is $\left(\frac{-2 + 3}{2}, \frac{3 - 2}{2}\right) = \left(\frac{1}{2}, \frac{1}{2}\right)$. The diagonals have the same midpoint so they bisect each other. Thus, *ABCD* is a parallelogram.

$AC = \sqrt{(-3 - 4)^2 + (0 - 1)^2} = \sqrt{49 + 1}$ or $\sqrt{50}$

$BD = \sqrt{(-2 - 3)^2 + (3 - (-2))^2} = \sqrt{25 + 25}$ or $\sqrt{50}$

The diagonals are congruent. *ABCD* is a parallelogram with diagonals that bisect each other, so it is a rectangle.

Exercises

Determine whether *ABCD* is a rectangle given each set of vertices. Justify your answer.

1. $A(-3, 1)$, $B(-3, 3)$, $C(3, 3)$, $D(3, 1)$

2. $A(-3, 0)$, $B(-2, 3)$, $C(4, 5)$, $D(3, 2)$

3. $A(-3, 0)$, $B(-2, 2)$, $C(3, 0)$, $D(2, -2)$

4. $A(-1, 0)$, $B(0, 2)$, $C(4, 0)$, $D(3, -2)$

5. $A(-1, -5)$, $B(-3, 0)$, $C(2, 2)$, $D(4, -3)$

6. $A(-1, -1)$, $B(0, 2)$, $C(4, 3)$, $D(3, 0)$

7. A parallelogram has vertices $R(-3, -1)$, $S(-1, 2)$, and $T(5, -2)$. Find the coordinates of U so that *RSTU* is a rectangle.

NAME ______________________ DATE __________ PERIOD _____

8-5 Study Guide and Intervention

Rhombi and Squares

Properties of Rhombi A **rhombus** is a quadrilateral with four congruent sides. Opposite sides are congruent, so a rhombus is also a parallelogram and has all of the properties of a parallelogram. Rhombi also have the following properties.

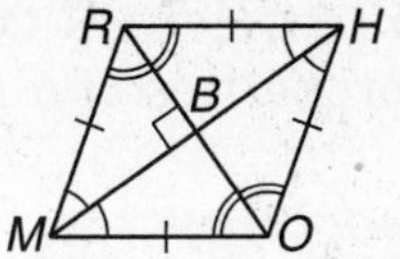

The diagonals are perpendicular.	$\overline{MH} \perp \overline{RO}$
Each diagonal bisects a pair of opposite angles.	$\overline{MH}$ bisects $\angle RMO$ and $\angle RHO$. $\overline{RO}$ bisects $\angle MRH$ and $\angle MOH$.
If the diagonals of a parallelogram are perpendicular, then the figure is a rhombus.	If $RHOM$ is a parallelogram and $\overline{RO} \perp \overline{MH}$, then $RHOM$ is a rhombus.

Example **In rhombus *ABCD*, $m\angle BAC = 32$. Find the measure of each numbered angle.**

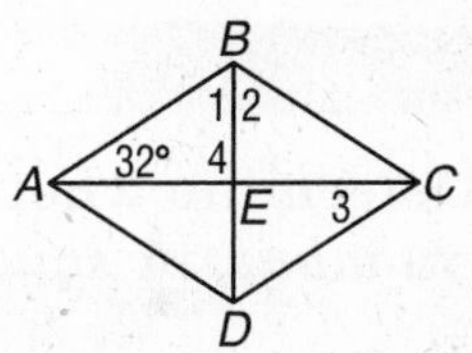

ABCD is a rhombus, so the diagonals are perpendicular and $\triangle ABE$ is a right triangle. Thus $m\angle 4 = 90$ and $m\angle 1 = 90 - 32$ or 58. The diagonals in a rhombus bisect the vertex angles, so $m\angle 1 = m\angle 2$. Thus, $m\angle 2 = 58$.

A rhombus is a parallelogram, so the opposite sides are parallel. $\angle BAC$ and $\angle 3$ are alternate interior angles for parallel lines, so $m\angle 3 = 32$.

Exercises

***ABCD* is a rhombus.**

1. If $m\angle ABD = 60$, find $m\angle BDC$.
2. If $AE = 8$, find AC.
3. If $AB = 26$ and $BD = 20$, find AE.
4. Find $m\angle CEB$.
5. If $m\angle CBD = 58$, find $m\angle ACB$.
6. If $AE = 3x - 1$ and $AC = 16$, find x.
7. If $m\angle CDB = 6y$ and $m\angle ACB = 2y + 10$, find y.
8. If $AD = 2x + 4$ and $CD = 4x - 4$, find x.
9. **a.** What is the midpoint of $\overline{FH}$?

   b. What is the midpoint of $\overline{GJ}$?

   c. What kind of figure is *FGHJ*? Explain.

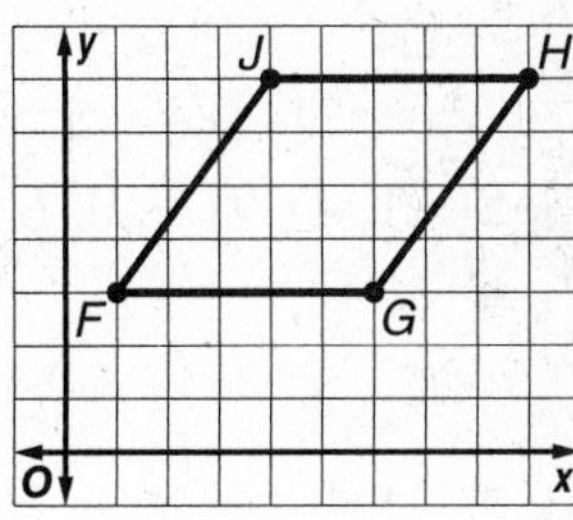

   d. What is the slope of $\overline{FH}$?

   e. What is the slope of $\overline{GJ}$?

   f. Based on parts **c**, **d**, and **e**, what kind of figure is *FGHJ*? Explain.

Lesson 8-5

8-5 Study Guide and Intervention *(continued)*

Rhombi and Squares

Properties of Squares A square has all the properties of a rhombus and all the properties of a rectangle.

Example **Find the measure of each numbered angle of square *ABCD*.**

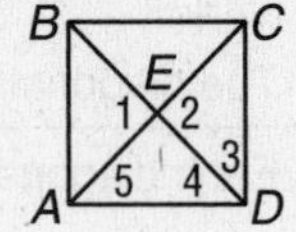

Using properties of rhombi and rectangles, the diagonals are perpendicular and congruent. $\triangle ABE$ is a right triangle, so $m\angle 1 = m\angle 2 = 90$.

Each vertex angle is a right angle and the diagonals bisect the vertex angles, so $m\angle 3 = m\angle 4 = m\angle 5 = 45$.

Exercises

Determine whether the given vertices represent a *parallelogram, rectangle, rhombus,* or *square*. Explain your reasoning.

1. $A(0, 2)$, $B(2, 4)$, $C(4, 2)$, $D(2, 0)$

2. $D(-2, 1)$, $E(-1, 3)$, $F(3, 1)$, $G(2, -1)$

3. $A(-2, -1)$, $B(0, 2)$, $C(2, -1)$, $D(0, -4)$

4. $A(-3, 0)$, $B(-1, 3)$, $C(5, -1)$, $D(3, -4)$

5. $S(-1, 4)$, $T(3, 2)$, $U(1, -2)$, $V(-3, 0)$

6. $F(-1, 0)$, $G(1, 3)$, $H(4, 1)$, $I(2, -2)$

7. Square *RSTU* has vertices $R(-3, -1)$, $S(-1, 2)$, and $T(2, 0)$. Find the coordinates of vertex *U*.

8-6 Study Guide and Intervention

Trapezoids

Properties of Trapezoids A **trapezoid** is a quadrilateral with exactly one pair of parallel sides. The parallel sides are called **bases** and the nonparallel sides are called **legs**. If the legs are congruent, the trapezoid is an **isosceles trapezoid**. In an isosceles trapezoid both pairs of **base angles** are congruent.

STUR is an isosceles trapezoid.
$\overline{SR} \cong \overline{TU}$; $\angle R \cong \angle U$, $\angle S \cong \angle T$

Example **The vertices of *ABCD* are *A*(−3, −1), *B*(−1, 3), *C*(2, 3), and *D*(4, −1). Verify that *ABCD* is a trapezoid.**

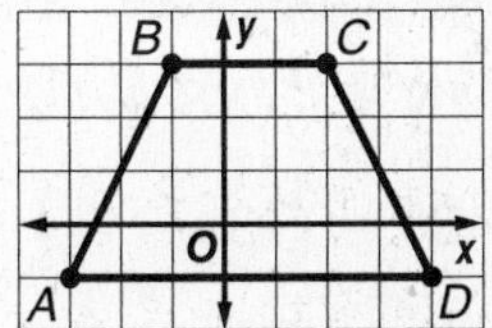

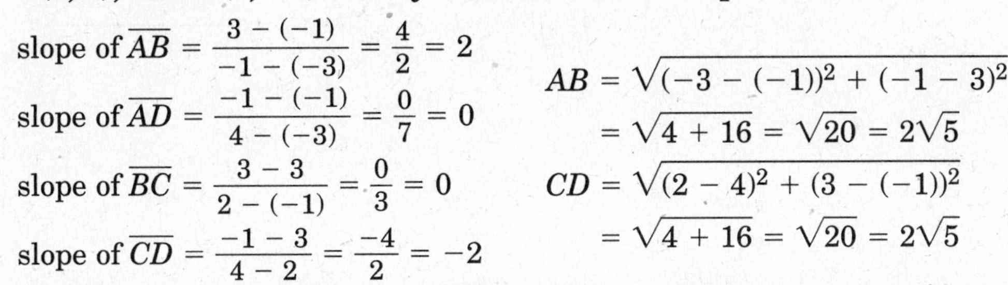

slope of $\overline{AB} = \dfrac{3-(-1)}{-1-(-3)} = \dfrac{4}{2} = 2$

slope of $\overline{AD} = \dfrac{-1-(-1)}{4-(-3)} = \dfrac{0}{7} = 0$

slope of $\overline{BC} = \dfrac{3-3}{2-(-1)} = \dfrac{0}{3} = 0$

slope of $\overline{CD} = \dfrac{-1-3}{4-2} = \dfrac{-4}{2} = -2$

$AB = \sqrt{(-3-(-1))^2 + (-1-3)^2}$
$= \sqrt{4+16} = \sqrt{20} = 2\sqrt{5}$

$CD = \sqrt{(2-4)^2 + (3-(-1))^2}$
$= \sqrt{4+16} = \sqrt{20} = 2\sqrt{5}$

Exactly two sides are parallel, $\overline{AD}$ and $\overline{BC}$, so *ABCD* is a trapezoid. *AB* = *CD*, so *ABCD* is an isosceles trapezoid.

Exercises

In Exercises 1–3, determine whether *ABCD* is a trapezoid. If so, determine whether it is an isosceles trapezoid. Explain.

1. *A*(−1, 1), *B*(2, 1), *C*(3, −2), and *D*(2, −2)

2. *A*(3, −3), *B*(−3, −3), *C*(−2, 3), and *D*(2, 3)

3. *A*(1, −4), *B*(−3, −3), *C*(−2, 3), and *D*(2, 2)

4. The vertices of an isosceles trapezoid are *R*(−2, 2), *S*(2, 2), *T*(4, −1), and *U*(−4, −1). Verify that the diagonals are congruent.

NAME ______________________ DATE ____________ PERIOD _____

8-6 Study Guide and Intervention *(continued)*

Trapezoids

Medians of Trapezoids The **median** of a trapezoid is the segment that joins the midpoints of the legs. It is parallel to the bases, and its length is one-half the sum of the lengths of the bases.

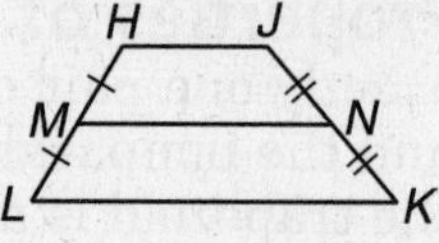

In trapezoid $HJKL$, $MN = \frac{1}{2}(HJ + LK)$.

Example $\overline{MN}$ **is the median of trapezoid** $RSTU$. **Find** x.

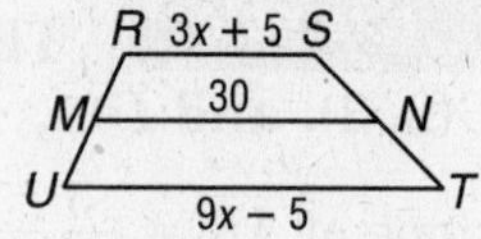

$MN = \frac{1}{2}(RS + UT)$

$30 = \frac{1}{2}(3x + 5 + 9x - 5)$

$30 = \frac{1}{2}(12x)$

$30 = 6x$

$5 = x$

Exercises

$\overline{MN}$ **is the median of trapezoid** $HJKL$. **Find each indicated value.**

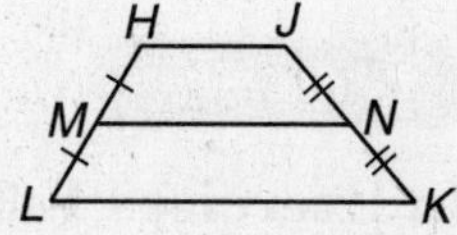

1. Find MN if $HJ = 32$ and $LK = 60$.

2. Find LK if $HJ = 18$ and $MN = 28$.

3. Find MN if $HJ + LK = 42$.

4. Find $m\angle LMN$ if $m\angle LHJ = 116$.

5. Find $m\angle JKL$ if $HJKL$ is isosceles and $m\angle HLK = 62$.

6. Find HJ if $MN = 5x + 6$, $HJ = 3x + 6$, and $LK = 8x$.

7. Find the length of the median of a trapezoid with vertices $A(-2, 2)$, $B(3, 3)$, $C(7, 0)$, and $D(-3, -2)$.

8-7 Study Guide and Intervention

Coordinate Proof with Quadrilaterals

Position Figures Coordinate proofs use properties of lines and segments to prove geometric properties. The first step in writing a coordinate proof is to place the figure on the coordinate plane in a convenient way. Use the following guidelines for placing a figure on the coordinate plane.

1. Use the origin as a vertex, so one set of coordinates is (0, 0), or use the origin as the center of the figure.
2. Place at least one side of the quadrilateral on an axis so you will have some zero coordinates.
3. Try to keep the quadrilateral in the first quadrant so you will have positive coordinates.
4. Use coordinates that make the computations as easy as possible. For example, use even numbers if you are going to be finding midpoints.

Example **Position and label a rectangle with sides a and b units long on the coordinate plane.**

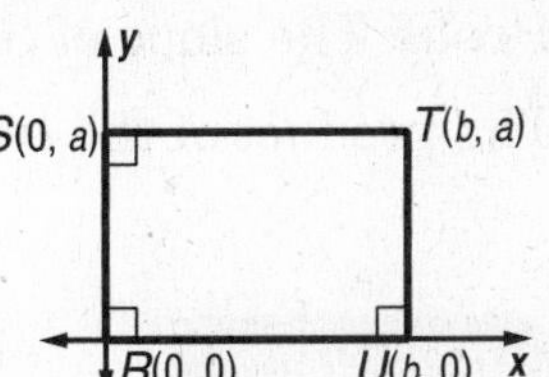

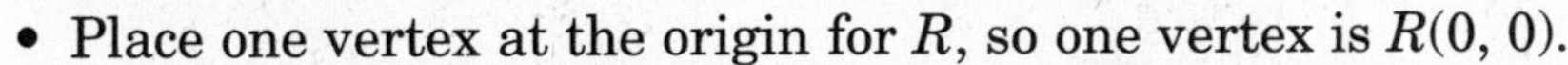

- Place one vertex at the origin for R, so one vertex is $R(0, 0)$.
- Place side $\overline{RU}$ along the x-axis and side $\overline{RS}$ along the y-axis, with the rectangle in the first quadrant.
- The sides are a and b units, so label two vertices $S(0, a)$ and $U(b, 0)$.
- Vertex T is b units right and a units up, so the fourth vertex is $T(b, a)$.

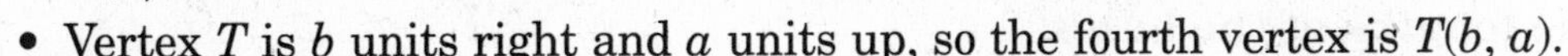

Exercises

Name the missing coordinates for each quadrilateral.

1. D(0, c), C(?, ?), A(0, 0), B(c, 0)

2. N(?, ?), P(a − b, c), M(0, 0), Q(a, 0)

3.

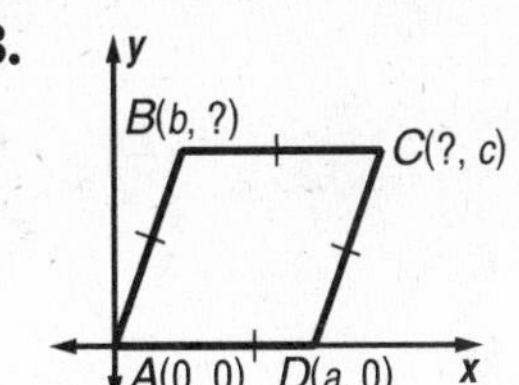

Position and label each quadrilateral on the coordinate plane.

4. square $STUV$ with side s units

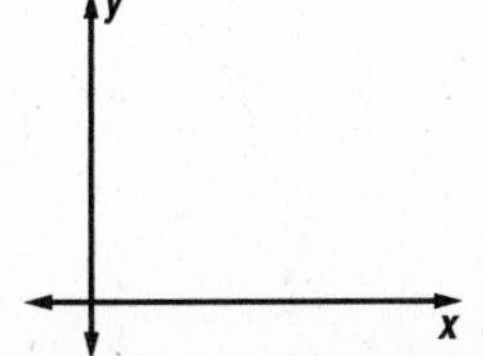

5. parallelogram $PQRS$ with congruent diagonals

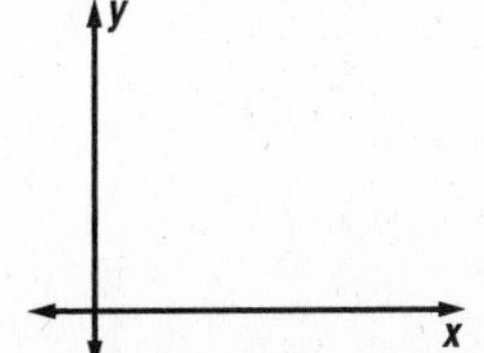

6. rectangle $ABCD$ with length twice the width

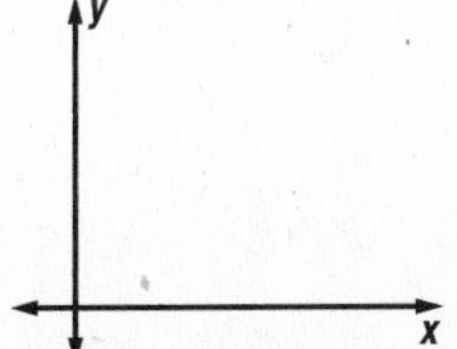

8-7 Study Guide and Intervention *(continued)*

Coordinate Proof With Quadrilaterals

Prove Theorems After a figure has been placed on the coordinate plane and labeled, a coordinate proof can be used to prove a theorem or verify a property. The Distance Formula, the Slope Formula, and the Midpoint Theorem are often used in a coordinate proof.

Example **Write a coordinate proof to show that the diagonals of a square are perpendicular.**

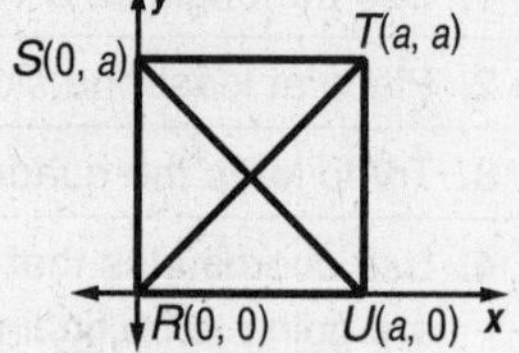

The first step is to position and label a square on the coordinate plane. Place it in the first quadrant, with one side on each axis. Label the vertices and draw the diagonals.

Given: square $RSTU$

Prove: $\overline{SU} \perp \overline{RT}$

Proof: The slope of $\overline{SU}$ is $\frac{0 - a}{a - 0} = -1$, and the slope of $\overline{RT}$ is $\frac{a - 0}{a - 0} = 1$.
The product of the two slopes is -1, so $\overline{SU} \perp \overline{RT}$.

Exercise

Write a coordinate proof to show that the length of the median of a trapezoid is half the sum of the lengths of the bases.

9-1 Study Guide and Intervention

Reflections

Draw Reflections The transformation called a **reflection** is a flip of a figure in a point, a line, or a plane. The new figure is the **image** and the original figure is the **preimage**. The preimage and image are congruent, so a reflection is a **congruence transformation** or **isometry**.

Example 1 **Construct the image of quadrilateral *ABCD* under a reflection in line *m*.**

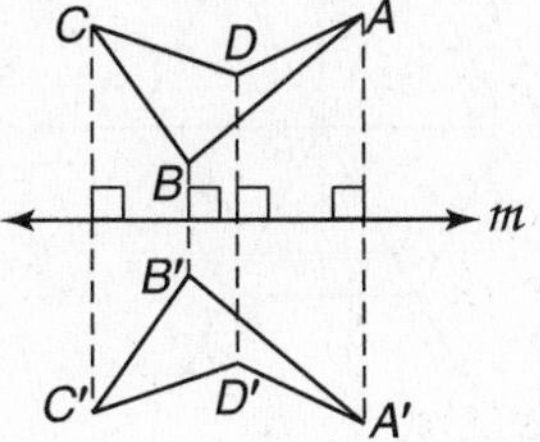

Draw a perpendicular from each vertex of the quadrilateral to *m*. Find vertices *A'*, *B'*, *C'*, and *D'* that are the same distance from *m* on the other side of *m*. The image is *A'B'C'D'*.

Example 2 **Quadrilateral *DEFG* has vertices $D(-2, 3)$, $E(4, 4)$, $F(3, -2)$, and $G(-3, -1)$. Find the image under reflection in the *x*-axis.**

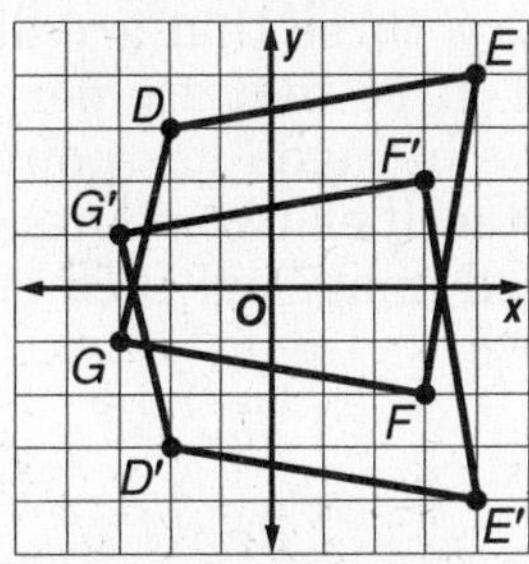

To find an image for a reflection in the *x*-axis, use the same *x*-coordinate and multiply the *y*-coordinate by -1. In symbols, $(a, b) \rightarrow (a, -b)$. The new coordinates are $D'(-2, -3)$, $E'(4, -4)$, $F'(3, 2)$, and $G'(-3, 1)$. The image is $D'E'F'G'$.

In Example 2, the notation $(a, b) \rightarrow (a, -b)$ represents a reflection in the *x*-axis. Here are three other common reflections in the coordinate plane.

- in the *y*-axis: $(a, b) \rightarrow (-a, b)$
- in the line $y = x$: $(a, b) \rightarrow (b, a)$
- in the origin: $(a, b) \rightarrow (-a, -b)$

Exercises

Draw the image of each figure under a reflection in line *m*.

1.

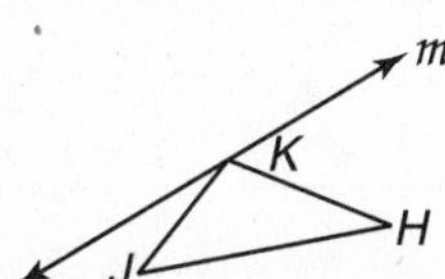

2.

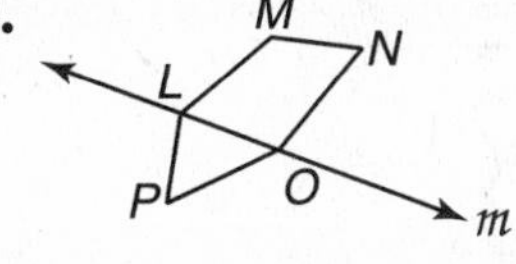

3.

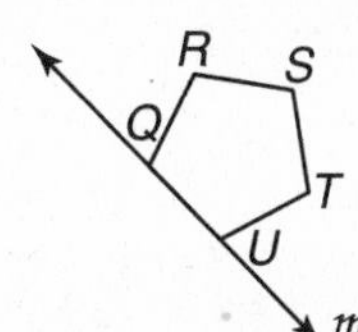

Graph each figure and its image under the given reflection.

4. $\triangle DEF$ with $D(-2, -1)$, $E(-1, 3)$, $F(3, -1)$ in the *x*-axis

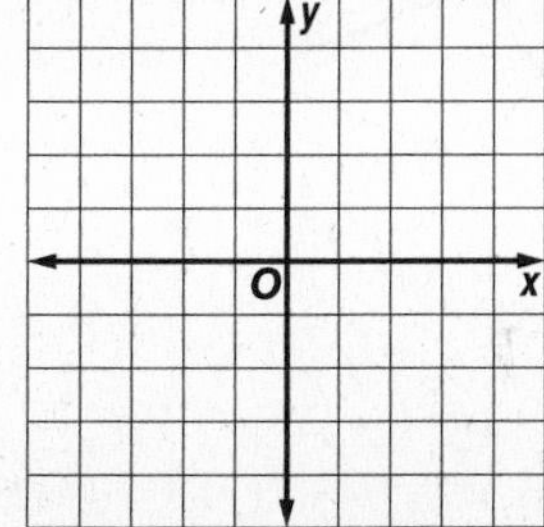

5. *ABCD* with $A(1, 4)$, $B(3, 2)$, $C(2, -2)$, $D(-3, 1)$ in the *y*-axis

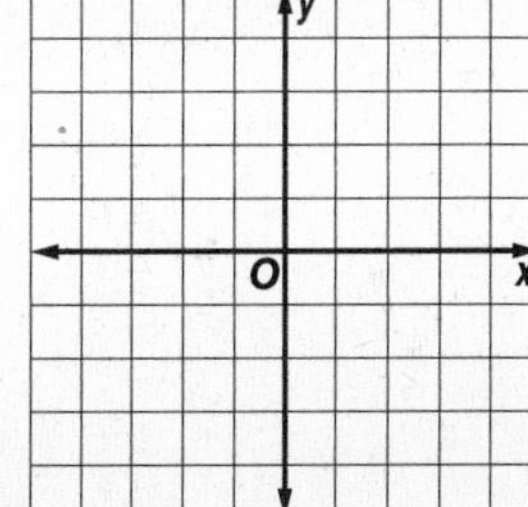

NAME ______________________________ DATE ____________ PERIOD ____

9-1 Study Guide and Intervention *(continued)*

Reflections

Lines and Points of Symmetry If a figure has a **line of symmetry**, then it can be folded along that line so that the two halves match. If a figure has a **point of symmetry**, it is the midpoint of all segments between the preimage and image points.

Example **Determine how many lines of symmetry a regular hexagon has. Then determine whether a regular hexagon has point symmetry.**

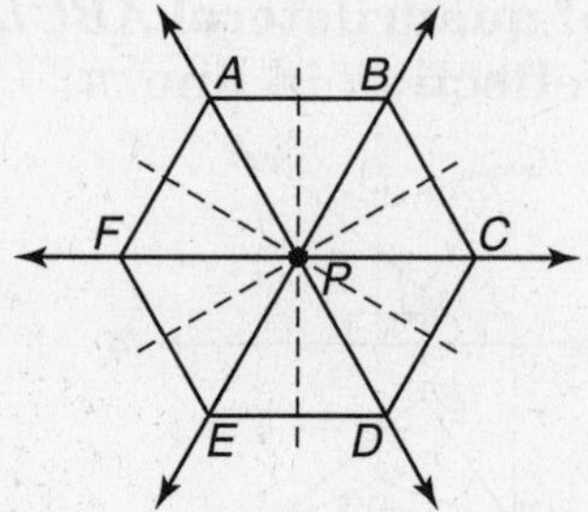

There are six lines of symmetry, three that are diagonals through opposite vertices and three that are perpendicular bisectors of opposite sides. The hexagon has point symmetry because any line through P identifies two points on the hexagon that can be considered images of each other.

Exercises

Determine how many lines of symmetry each figure has. Then determine whether the figure has point symmetry.

1.

2.

3.

4.

5.

6.

7.

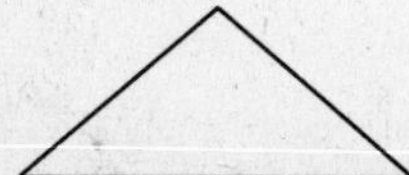

8.

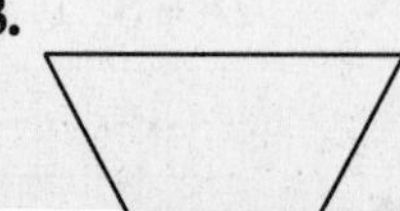

9.

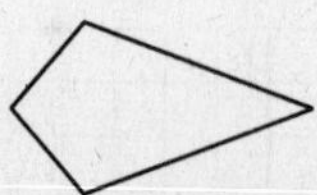

© Glencoe/McGraw-Hill

9-2 Study Guide and Intervention

Translations

Translations Using Coordinates A transformation called a **translation** slides a figure in a given direction. In the coordinate plane, a translation moves every preimage point $P(x, y)$ to an image point $P(x + a, y + b)$ for fixed values a and b. In words, a translation shifts a figure a units horizontally and b units vertically; in symbols, $(x, y) \rightarrow (x + a, y + b)$.

Example **Rectangle *RECT* has vertices $R(-2, -1)$, $E(-2, 2)$, $C(3, 2)$, and $T(3, -1)$. Graph *RECT* and its image for the translation $(x, y) \rightarrow (x + 2, y - 1)$.**

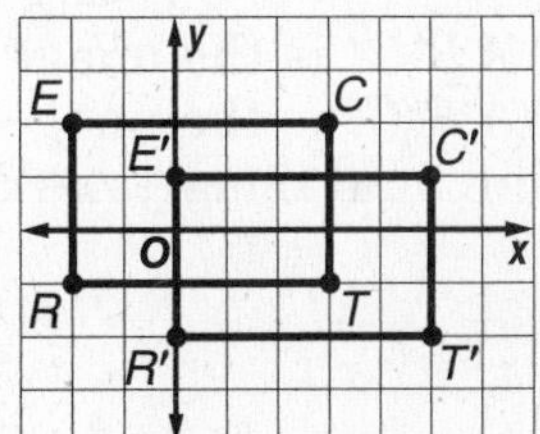

The translation moves every point of the preimage right 2 units and down 1 unit.

$(x, y) \rightarrow (x + 2, y - 1)$

$R(-2, -1) \rightarrow R'(-2 + 2, -1 - 1)$ or $R'(0, -2)$

$E(-2, 2) \rightarrow E'(-2 + 2, 2 - 1)$ or $E'(0, 1)$

$C(3, 2) \rightarrow C'(3 + 2, 2 - 1)$ or $C'(5, 1)$

$T(3, -1) \rightarrow T'(3 + 2, -1 - 1)$ or $T'(5, -2)$

Exercises

Graph each figure and its image under the given translation.

1. $\overline{PQ}$ with endpoints $P(-1, 3)$ and $Q(2, 2)$ under the translation left 2 units and up 1 unit

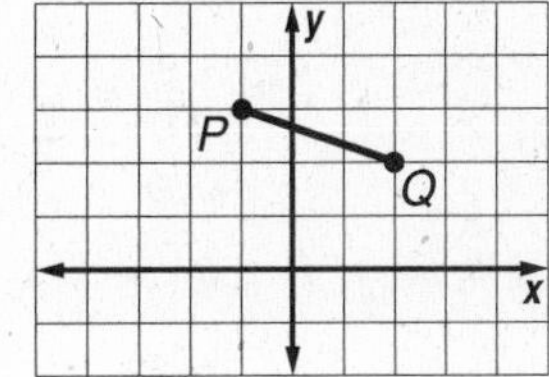

2. $\triangle PQR$ with vertices $P(-2, -4)$, $Q(-1, 2)$, and $R(2, 1)$ under the translation right 2 units and down 2 units

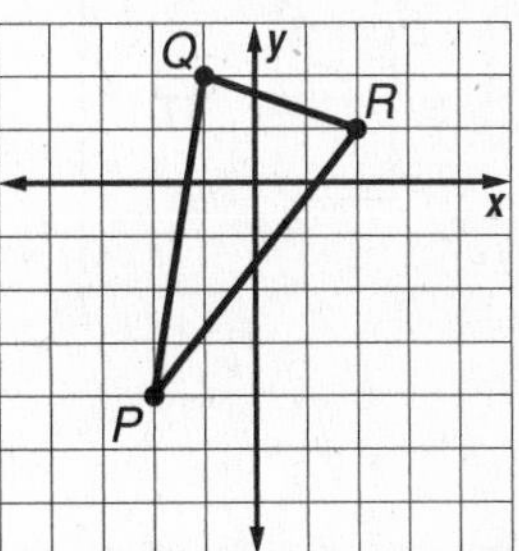

3. square *SQUR* with vertices $S(0, 2)$, $Q(3, 1)$, $U(2, -2)$, and $R(-1, -1)$ under the translation right 3 units and up 1 unit

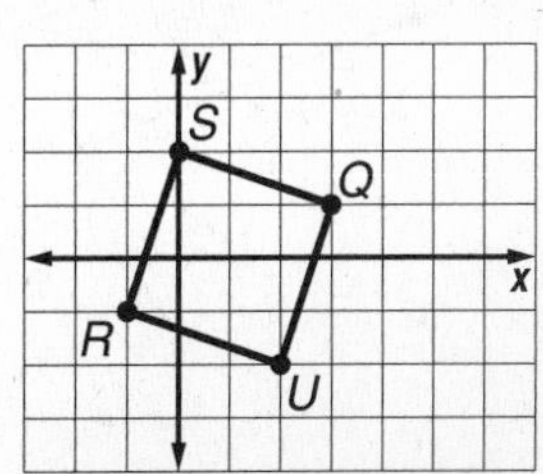

Lesson 9-2

9-2 Study Guide and Intervention *(continued)*

Translations

Translations by Repeated Reflections Another way to find the image of a translation is to reflect the figure twice in parallel lines. This kind of translation is called a **composite of reflections.**

Example **In the figure, $m \parallel n$. Find the translation image of $\triangle ABC$.**

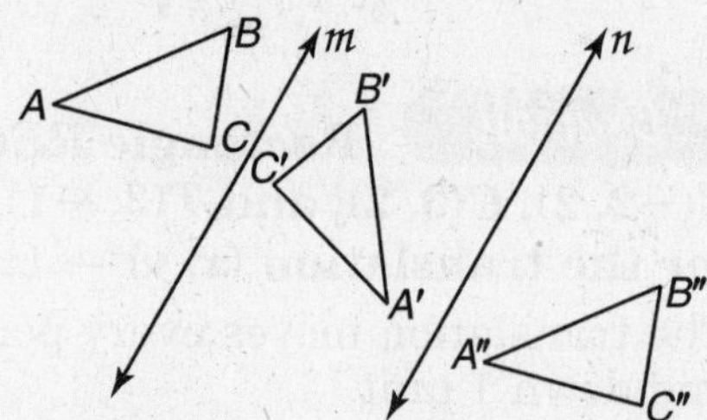

$\triangle A'B'C'$ is the image of $\triangle ABC$ reflected in line m.
$\triangle A''B''C''$ is the image of $\triangle A'B'C'$ reflected in line n.
The final image, $\triangle A''B''C''$, is a translation of $\triangle ABC$.

Exercises

In each figure, $m \parallel n$. Find the translation image of each figure by reflecting it in line m and then in line n.

1.

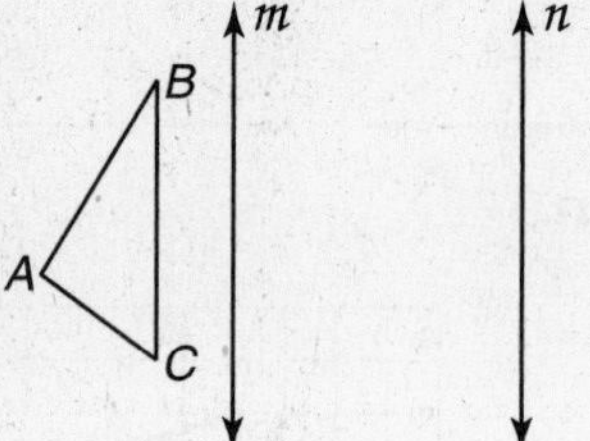

2.

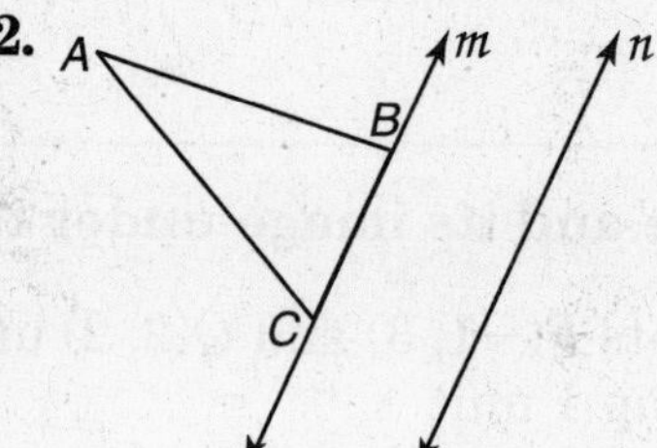

3.

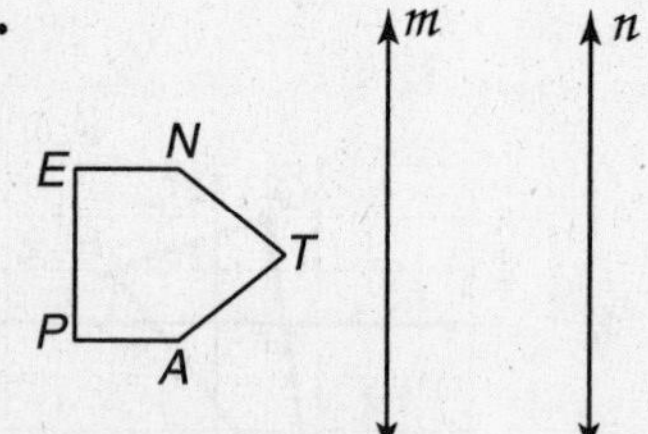

4.

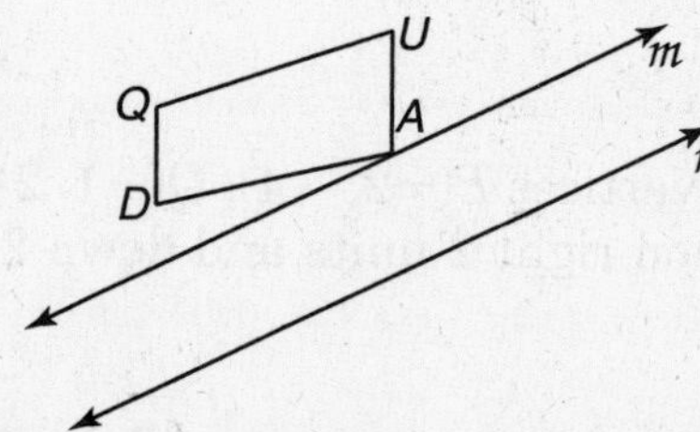

5.

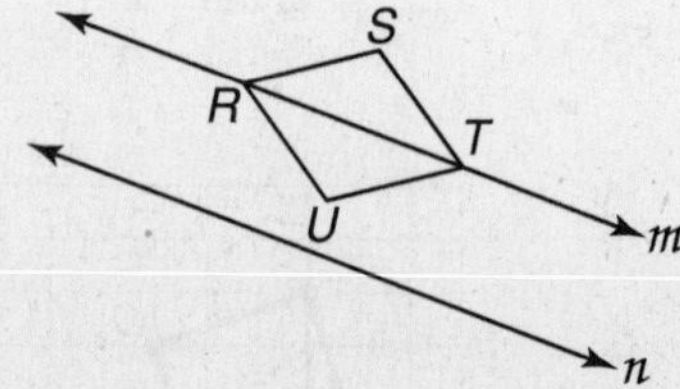

6.

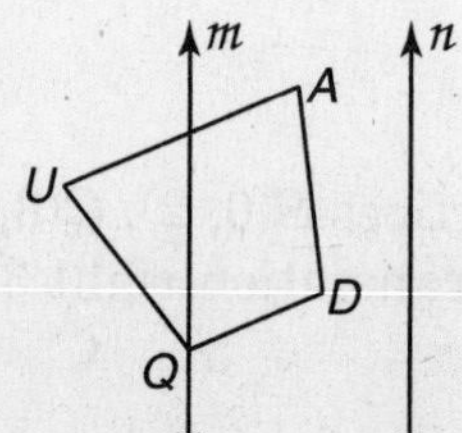

9-3 Study Guide and Intervention

Rotations

Draw Rotations A transformation called a **rotation** turns a figure through a specified angle about a fixed point called the **center of rotation**. To find the image of a rotation, one way is to use a protractor. Another way is to reflect a figure twice, in two intersecting lines.

Example 1 **$\triangle ABC$ has vertices $A(2, 1)$, $B(3, 4)$, and $C(5, 1)$. Draw the image of $\triangle ABC$ under a rotation of 90° counterclockwise about the origin.**

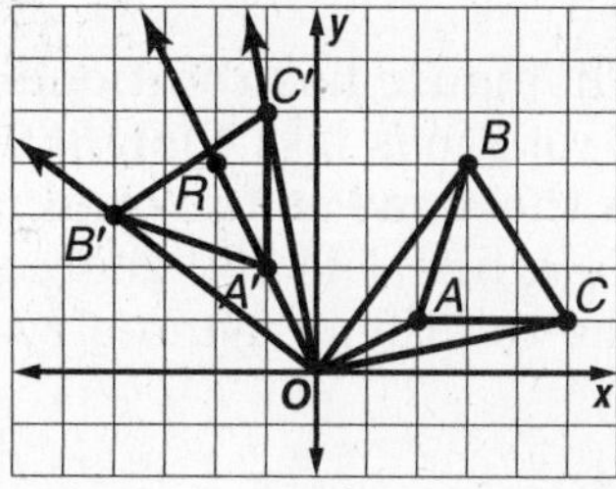

- First draw $\triangle ABC$. Then draw a segment from O, the origin, to point A.
- Use a protractor to measure 90° counterclockwise with $\overline{OA}$ as one side.
- Draw $\overrightarrow{OR}$.
- Use a compass to copy $\overline{OA}$ onto $\overrightarrow{OR}$. Name the segment $\overline{OA'}$.
- Repeat with segments from the origin to points B and C.

Example 2 **Find the image of $\triangle ABC$ under reflection in lines m and n.**

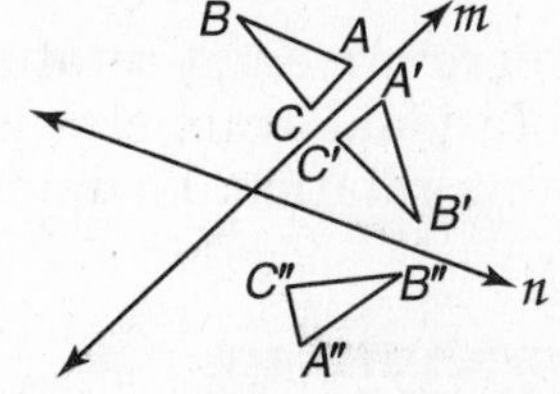

First reflect $\triangle ABC$ in line m. Label the image $\triangle A'B'C'$.

Reflect $\triangle A'B'C'$ in line n. Label the image $\triangle A''B''C''$.

$\triangle A''B''C''$ is a rotation of $\triangle ABC$. The center of rotation is the intersection of lines m and n. The angle of rotation is twice the measure of the acute angle formed by m and n.

Exercises

Draw the rotation image of each figure 90° in the given direction about the center point and label the coordinates.

1. $\overline{PQ}$ with endpoints $P(-1, -2)$ and $Q(1, 3)$ counterclockwise about the origin

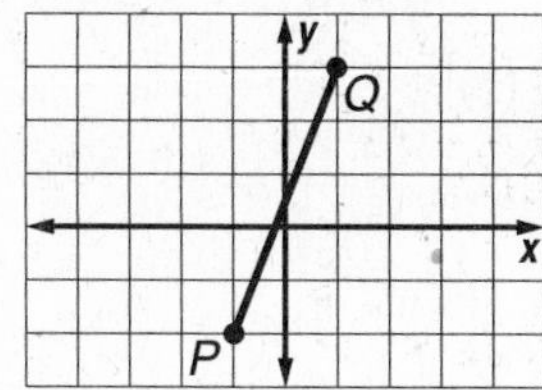

2. $\triangle PQR$ with vertices $P(-2, -3)$, $Q(2, -1)$, and $R(3, 2)$ clockwise about the point $T(1, 1)$

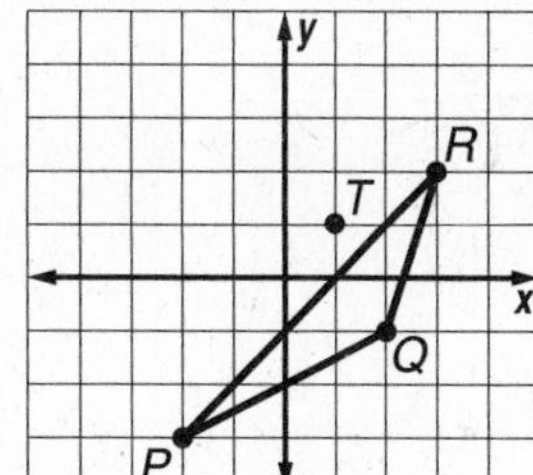

Find the rotation image of each figure by reflecting it in line m and then in line n.

3.

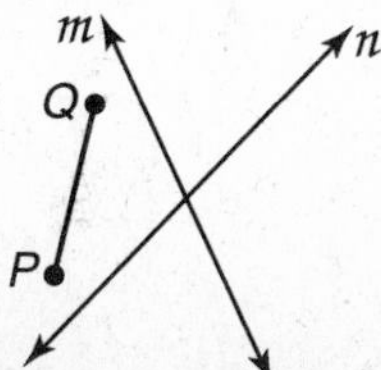

4.

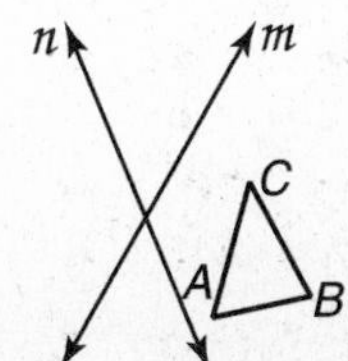

© Glencoe/McGraw-Hill

9-3 Study Guide and Intervention *(continued)*

Rotations

Rotational Symmetry When the figure at the right is rotated about point P by 120° or 240°, the image looks like the preimage. The figure has **rotational symmetry**, which means it can be rotated less than 360° about a point and the preimage and image appear to be the same.

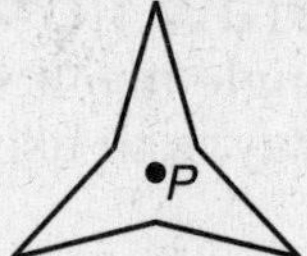

The figure has rotational symmetry of **order** 3 because there are 3 rotations less than 360° (0°, 120°, 240°) that produce an image that is the same as the original. The **magnitude** of the rotational symmetry for a figure is 360 degrees divided by the order. For the figure above, the rotational symmetry has magnitude 120 degrees.

Example **Identify the order and magnitude of the rotational symmetry of the design at the right.**

The design has rotational symmetry about the center point for rotations of 0°, 45°, 90°, 135°, 180°, 225°, 270°, and 315°.

There are eight rotations less than 360 degrees, so the order of its rotational symmetry is 8. The quotient 360 ÷ 8 is 45, so the magnitude of its rotational symmetry is 45 degrees.

Exercises

Identify the order and magnitude of the rotational symmetry of each figure.

1. a square

2. a regular 40-gon

3.

4.

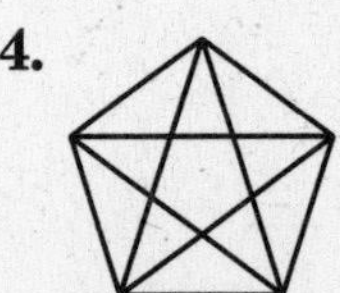

5.

6.

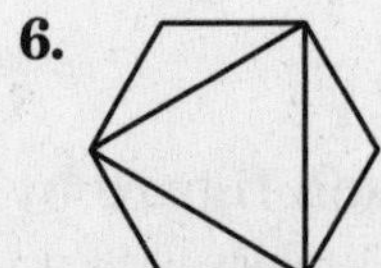

9-4 Study Guide and Intervention

Tessellations

Regular Tessellations A pattern that covers a plane with repeating copies of one or more figures so that there are no overlapping or empty spaces is a **tessellation.** A **regular tessellation** uses only one type of regular polygon. In a tessellation, the sum of the measures of the angles of the polygons surrounding a vertex is 360. If a regular polygon has an interior angle that is a factor of 360, then the polygon will tessellate.

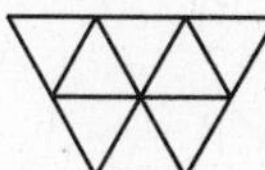
regular tessellation

tessellation

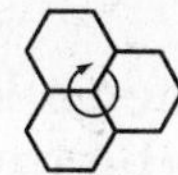
Copies of a regular hexagon can form a tessellation.

Copies of a regular pentagon cannot form a tessellation.

Example **Determine whether a regular 16-gon tessellates the plane. Explain.**

If $m\angle 1$ is the measure of one interior angle of a regular polygon, then a formula for $m\angle 1$ is $m\angle 1 = \frac{180(n-2)}{n}$. Use the formula with $n = 16$.

$$m\angle 1 = \frac{180(n-2)}{n}$$
$$= \frac{180(16-2)}{16}$$
$$= 157.5$$

The value 157.5 is not a factor of 360, so the 16-gon will not tessellate.

Exercises

Determine whether each polygon tessellates the plane. If so, draw a sample figure.

1. scalene right triangle

2. isosceles trapezoid

Determine whether each regular polygon tessellates the plane. Explain.

3. square

4. 20-gon

5. septagon

6. 15-gon

7. octagon

8. pentagon

Lesson 9-4

9-4 Study Guide and Intervention *(continued)*

Tessellations

Tessellations with Specific Attributes A tessellation pattern can contain any type of polygon. If the arrangement of shapes and angles at each vertex in the tessellation is the same, the tessellation is **uniform**. A **semi-regular tessellation** is a uniform tessellation that contains two or more regular polygons.

Example **Determine whether a kite will tessellate the plane. If so, describe the tessellation as *uniform*, *regular*, *semi-regular*, or *not uniform*.**

A kite will tessellate the plane. At each vertex the sum of the measures is $a + b + b + c$, which is 360. The tessellation is uniform.

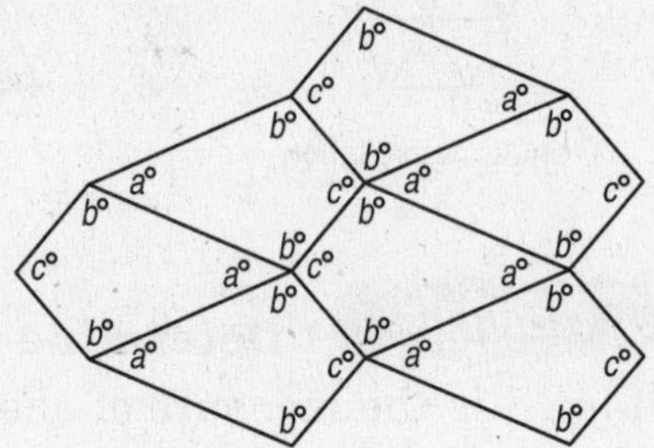

Exercises

Determine whether a semi-regular tessellation can be created from each set of figures. If so, sketch the tessellation. Assume that each figure has a side length of 1 unit.

1. rhombus, equilateral triangle, and octagon

2. square and equilateral triangle

Determine whether each polygon tessellates the plane. If so, describe the tessellation as *uniform*, *not uniform*, *regular*, or *semi-regular*.

3. rectangle

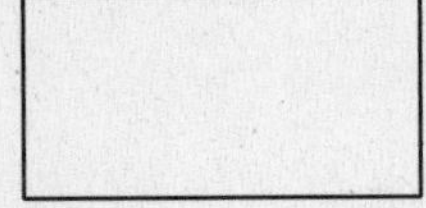

4. hexagon and square

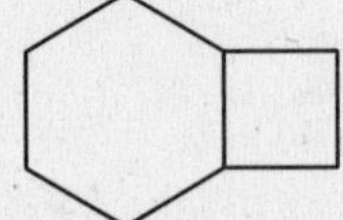

9-5 Study Guide and Intervention

Dilations

Classify Dilations A **dilation** is a transformation in which the image may be a different size than the preimage. A dilation requires a center point and a scale factor, r.

Let r represent the scale factor of a dilation.
If $\|r\| > 1$, then the dilation is an enlargement.
If $\|r\| = 1$, then the dilation is a congruence transformation.
If $0 < \|r\| < 1$, then the dilation is a reduction.

Example **Draw the dilation image of $\triangle ABC$ with center O and $r = 2$.**

Draw $\overrightarrow{OA}$, $\overrightarrow{OB}$, and $\overrightarrow{OC}$. Label points A', B', and C' so that $OA' = 2(OA)$, $OB' = 2(OB)$, and $OC' = 2(OC)$. $\triangle A'B'C'$ is a dilation of $\triangle ABC$.

O•

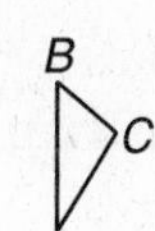

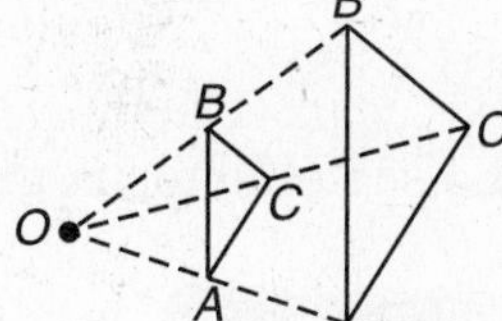

Exercises

Draw the dilation image of each figure with center C and the given scale factor. Describe each transformation as an *enlargement*, *congruence*, or *reduction*.

1. $r = 2$

C•

B

A

2. $r = \frac{1}{2}$

•C

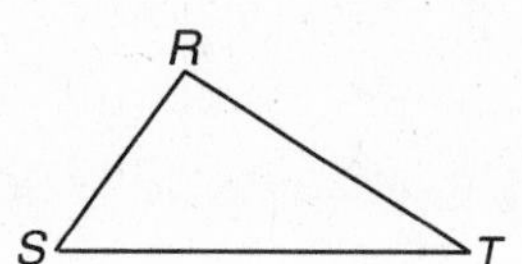

3. $r = 1$

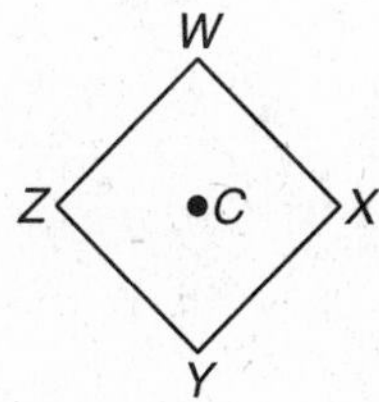

4. $r = 3$

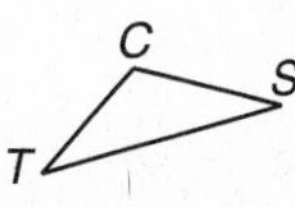

5. $r = \frac{2}{3}$

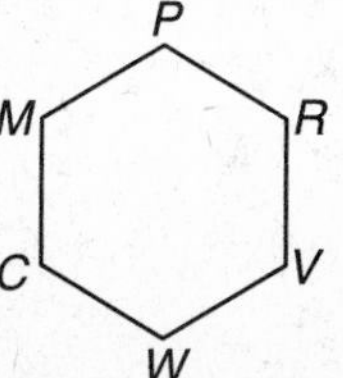

6. $r = 1$

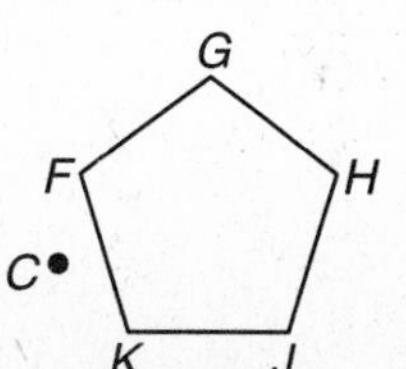

9-5 Study Guide and Intervention *(continued)*

Dilations

Identify the Scale Factor If you know corresponding measurements for a preimage and its dilation image, you can find the scale factor.

Example **Determine the scale factor for the dilation of $\overline{XY}$ to $\overline{AB}$. Determine whether the dilation is an *enlargement*, *reduction*, or *congruence transformation*.**

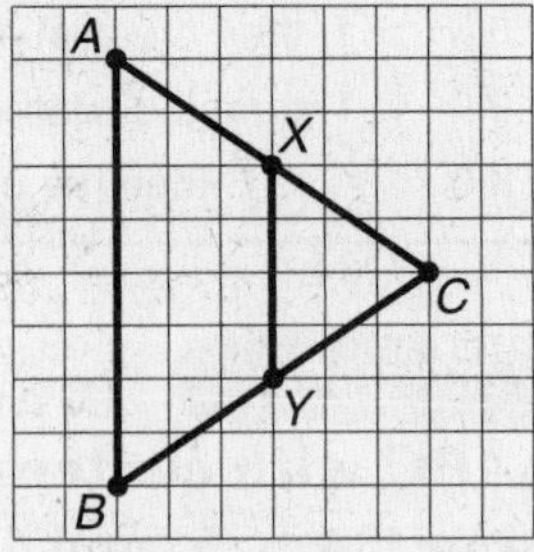

$$\text{scale factor} = \frac{\text{image length}}{\text{preimage length}} = \frac{8 \text{ units}}{4 \text{ units}} = 2$$

The scale factor is greater than 1, so the dilation is an enlargement.

Exercises

Determine the scale factor for each dilation with center *C*. Determine whether the dilation is an *enlargement*, *reduction*, or *congruence transformation*.

1. *CGHJ* is a dilation image of *CDEF*.

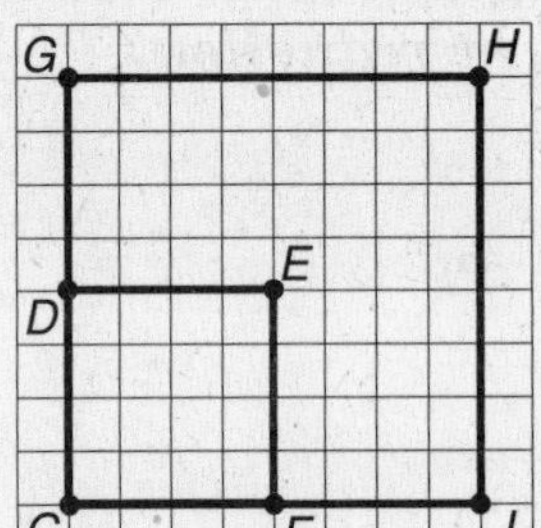

2. △*CKL* is a dilation image of △*CKL*.

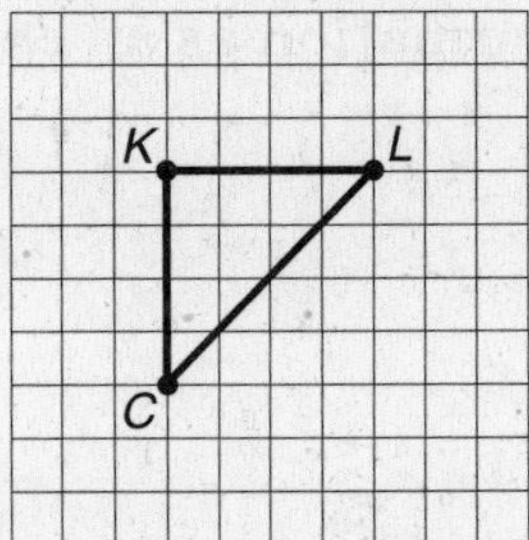

3. *STUVWX* is a dilation image of *MNOPQR*.

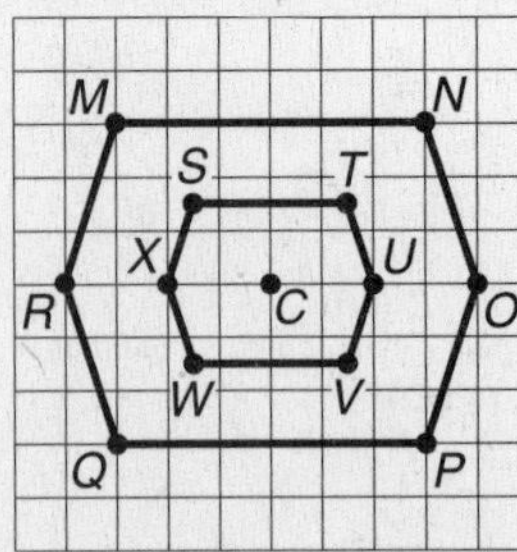

4. △*CPQ* is a dilation image of △*CYZ*.

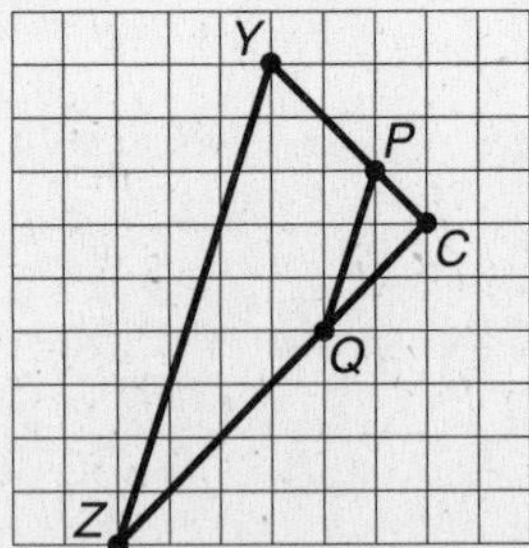

5. △*EFG* is a dilation image of △*ABC*.

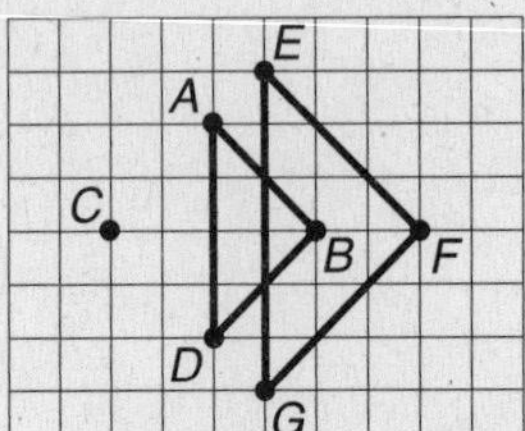

6. △*HJK* is a dilation image of △*HJK*.

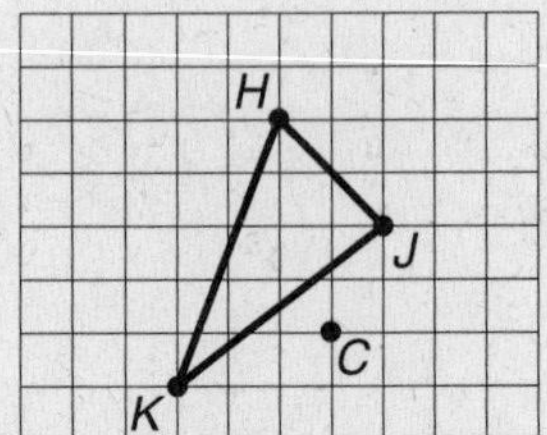

Study Guide and Intervention

Vectors

Magnitude and Direction A vector is a directed segment representing a quantity that has both **magnitude**, or length, and **direction**. For example, the speed and direction of an airplane can be represented by a vector. In symbols, a vector is written as $\overrightarrow{AB}$, where A is the initial point and B is the endpoint, or as $\vec{\mathbf{v}}$.

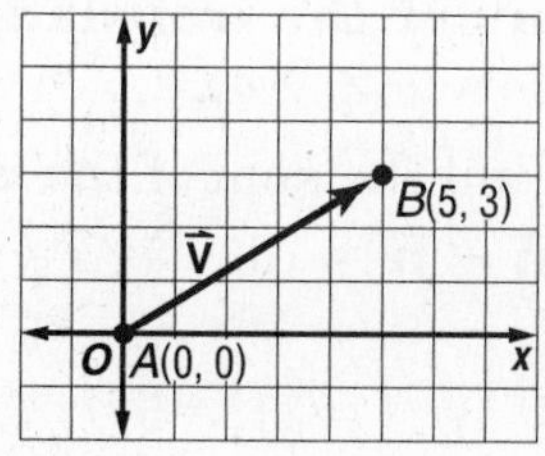

A vector in **standard position** has its initial point at (0, 0) and can be represented by the ordered pair for point B. The vector at the right can be expressed as $\vec{\mathbf{v}} = \langle 5, 3 \rangle$.

You can use the Distance Formula to find the magnitude $|\overrightarrow{AB}|$ of a vector. You can describe the direction of a vector by measuring the angle that the vector forms with the positive x-axis or with any other horizontal line.

Example **Find the magnitude and direction of $\overrightarrow{AB}$ for $A(5, 2)$ and $B(8, 7)$.**

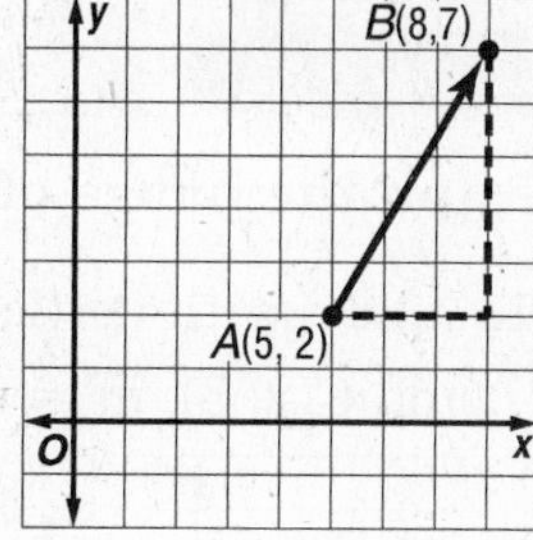

Find the magnitude.

$$|AB| = \sqrt{(x_2 - x_1)^2 + (y_2 - y_1)^2}$$
$$= \sqrt{(8 - 5)^2 + (7 - 2)^2}$$
$$= \sqrt{34} \text{ or about 5.8 units}$$

To find the direction, use the tangent ratio.

$\tan A = \frac{5}{3}$ The tangent ratio is opposite over adjacent.

$m\angle A \approx 59.0$ Use a calculator.

The magnitude of the vector is about 5.8 units and its direction is 59°.

Exercises

Find the magnitude and direction of $\overrightarrow{AB}$ for the given coordinates. Round to the nearest tenth.

1. $A(3, 1)$, $B(-2, 3)$

2. $A(0, 0)$, $B(-2, 1)$

3. $A(0, 1)$, $B(3, 5)$

4. $A(-2, 2)$, $B(3, 1)$

5. $A(3, 4)$, $B(0, 0)$

6. $A(4, 2)$, $B(0, 3)$

9-6 Study Guide and Intervention *(continued)*

Vectors

Translations with Vectors Recall that the transformation $(a, b) \rightarrow (a + 2, b - 3)$ represents a translation right 2 units and down 3 units. The vector $\langle 2, -3\rangle$ is another way to describe that translation. Also, two vectors can be added: $\langle a, b\rangle + \langle c, d\rangle = \langle a + c, b + d\rangle$. The sum of two vectors is called the **resultant**.

Example **Graph the image of parallelogram *RSTU* under the translation by the vectors $\vec{m} = \langle 3, -1\rangle$ and $\vec{n} = \langle -2, -4\rangle$.**

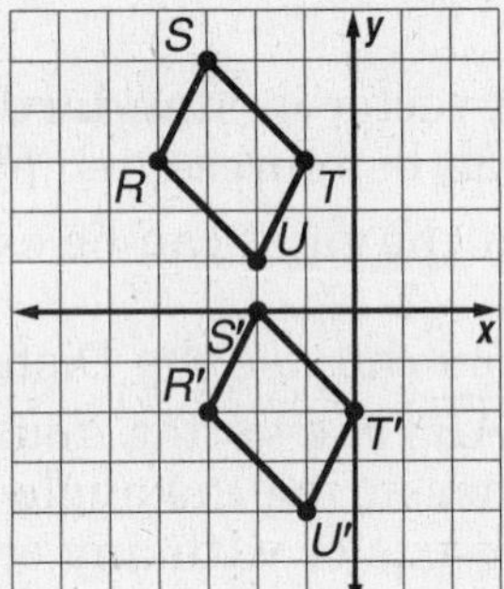

Find the sum of the vectors.

$\vec{m} + \vec{n} = \langle 3, -1\rangle + \langle -2, -4\rangle$
$= \langle 3 - 2, -1 - 4\rangle$
$= \langle 1, -5\rangle$

Translate each vertex of parallelogram *RSTU* right 1 unit and down 5 units.

Exercises

Graph the image of each figure under a translation by the given vector(s).

1. $\triangle ABC$ with vertices $A(-1, 2)$, $B(0, 0)$, and $C(2, 3)$; $\vec{m} = \langle 2, -3\rangle$

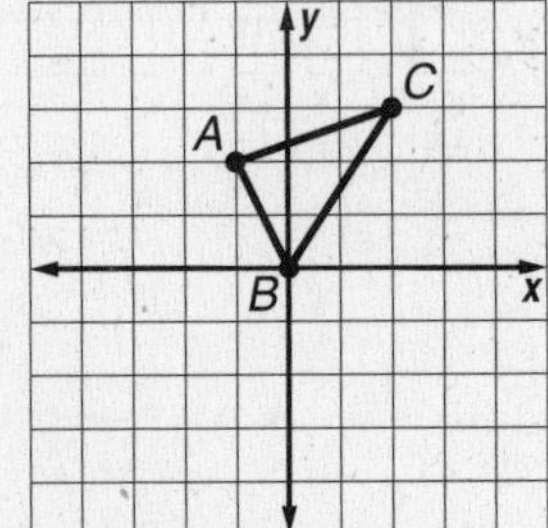

2. *ABCD* with vertices $A(-4, 1)$, $B(-2, 3)$, $C(1, 1)$, and $D(-1, -1)$; $\vec{n} = \langle 3 -3\rangle$

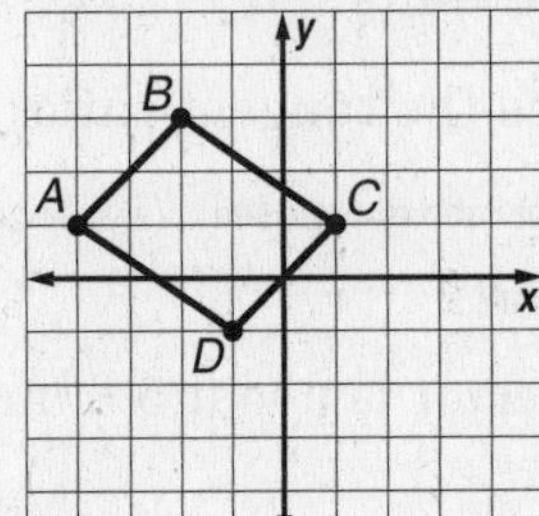

3. *ABCD* with vertices $A(-3, 3)$, $B(1, 3)$, $C(1, 1)$, and $D(-3, 1)$; the sum of $\vec{p} = \langle -2, 1\rangle$ and $\vec{q} = \langle 5, -4\rangle$

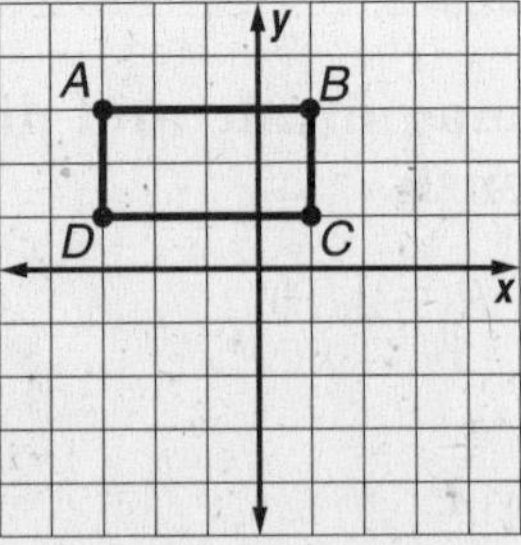

Given $\vec{m} = \langle 1, -2\rangle$ and $\vec{n} = \langle -3, -4\rangle$, represent each of the following as a single vector.

4. $\vec{m} + \vec{n}$

5. $\vec{n} - \vec{m}$

9-7 Study Guide and Intervention

Transformations with Matrices

Translations and Dilations A **vector** can be represented by the ordered pair $\langle x, y \rangle$ or by the **column matrix** $\begin{bmatrix} x \\ y \end{bmatrix}$. When the ordered pairs for all the vertices of a polygon are placed together, the resulting matrix is called the **vertex matrix** for the polygon.

For $\triangle ABC$ with $A(-2, 2)$, $B(2, 1)$, and $C(-1, -1)$, the vertex matrix for the triangle is $\begin{bmatrix} -2 & 2 & -1 \\ 2 & 1 & -1 \end{bmatrix}$.

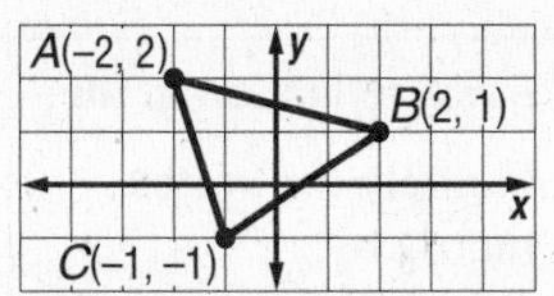

Example 1 **For $\triangle ABC$ above, use a matrix to find the coordinates of the vertices of the image of $\triangle ABC$ under the translation $(x, y) \rightarrow (x + 3, y - 1)$.**

To translate the figure 3 units to the right, add 3 to each x-coordinate. To translate the figure 1 unit down, add -1 to each y-coordinate.

Vertex Matrix of $\triangle ABC$		Translation Matrix		Vertex Matrix of $\triangle A'B'C'$
$\begin{bmatrix} -2 & 2 & -1 \\ 2 & 1 & -1 \end{bmatrix}$	$+$	$\begin{bmatrix} 3 & 3 & 3 \\ -1 & -1 & -1 \end{bmatrix}$	$=$	$\begin{bmatrix} 1 & 5 & 2 \\ 1 & 0 & -2 \end{bmatrix}$

The coordinates are $A'(1, 1)$, $B'(5, 0)$, and $C'(2, -2)$.

Example 2 **For $\triangle ABC$ above, use a matrix to find the coordinates of the vertices of the image of $\triangle ABC$ for a dilation centered at the origin with scale factor 3.**

Scale Factor		Vertex Matrix of $\triangle ABC$		Vertex Matrix of $\triangle A'B'C'$
3	$\cdot$	$\begin{bmatrix} -2 & 2 & -1 \\ 2 & 1 & -1 \end{bmatrix}$	$=$	$\begin{bmatrix} -6 & 6 & -3 \\ 6 & 3 & -3 \end{bmatrix}$

The coordinates are $A'(-6, 6)$, $B'(6, 3)$, and $C'(-3, -3)$.

Exercises

Use a matrix to find the coordinates of the vertices of the image of each figure under the given translations or dilations.

1. $\triangle ABC$ with $A(3, 1)$, $B(-2, 4)$, $C(-2, -1)$; $(x, y) \rightarrow (x - 1, y + 2)$

2. parallelogram $RSTU$ with $R(-4, -2)$, $S(-3, 1)$, $T(3, 4)$, $U(2, 1)$; $(x, y) \rightarrow (x - 4, y - 3)$

3. rectangle $PQRS$ with $P(4, 0)$, $Q(3, -3)$, $R(-3, -1)$, $S(-2, 2)$; $(x, y) \rightarrow (x - 2, y + 1)$

4. $\triangle ABC$ with $A(-2, -1)$, $B(-2, -3)$, $C(2, -1)$; dilation centered at the origin with scale factor 2

5. parallelogram $RSTU$ with $R(4, -2)$, $S(-4, -1)$, $T(-2, 3)$, $U(6, 2)$; dilation centered at the origin with scale factor 1.5

9-7 Study Guide and Intervention *(continued)*

Transformations with Matrices

Reflections and Rotations When you reflect an image, one way to find the coordinates of the reflected vertices is to multiply the vertex matrix of the object by a **reflection matrix**. To perform more than one reflection, multiply by one reflection matrix to find the first image. Then multiply by the second matrix to find the final image. The matrices for reflections in the axes, the origin, and the line $y = x$ are shown below.

For a reflection in the:	*x*-axis	*y*-axis	origin	line $y = x$
Multiply the vertex matrix by:	$\begin{bmatrix} 1 & 0 \\ 0 & -1 \end{bmatrix}$	$\begin{bmatrix} -1 & 0 \\ 0 & 1 \end{bmatrix}$	$\begin{bmatrix} -1 & 0 \\ 0 & -1 \end{bmatrix}$	$\begin{bmatrix} 0 & 1 \\ 1 & 0 \end{bmatrix}$

Example **$\triangle ABC$ has vertices $A(-2, 3)$, $B(1, 4)$, and $C(3, 0)$. Use a matrix to find the coordinates of the vertices of the image of $\triangle ABC$ after a reflection in the x-axis.**

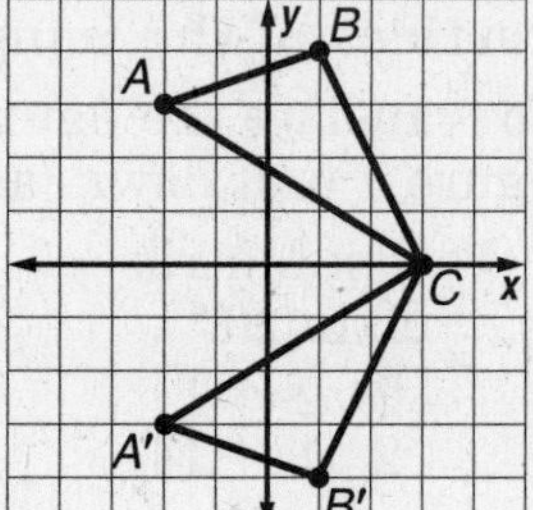

To reflect in the x-axis, multiply the vertex matrix of $\triangle ABC$ by the reflection matrix for the x-axis.

Reflection Matrix for x-axis · Vertex Matrix of $\triangle ABC$ = Vertex Matrix of $\triangle A'B'C'$

$$\begin{bmatrix} 1 & 0 \\ 0 & -1 \end{bmatrix} \cdot \begin{bmatrix} -2 & 1 & 3 \\ 3 & 4 & 0 \end{bmatrix} = \begin{bmatrix} -2 & 1 & 3 \\ -3 & -4 & 0 \end{bmatrix}$$

Exercises

Use a matrix to find the coordinates of the vertices of the image of each figure under the given reflection.

1. $\triangle ABC$ with $A(-3, 2)$, $B(-1, 3)$, $C(1, 0)$; reflection in the x-axis

2. $\triangle XYZ$ with $X(2, -1)$, $Y(4, -3)$, $Z(-2, 1)$; reflection in the y-axis

3. $\triangle ABC$ with $A(3, 4)$, $B(-1, 0)$, $C(-2, 4)$; reflection in the origin

4. parallelogram $RSTU$ with $R(-3, 2)$, $S(3, 2)$, $T(5, -1)$, $U(-1, -1)$; reflection in the line $y = x$

5. $\triangle ABC$ with $A(2, 3)$, $B(-1, 2)$, $C(1, -1)$; reflection in the origin, then reflection in the line $y = x$

6. parallelogram $RSTU$ with $R(0, 2)$, $S(4, 2)$, $T(3, -2)$, $U(-1, -2)$; reflection in the x-axis, then reflection in the y-axis

NAME ______________________________ DATE ____________ PERIOD _____

10-1 Study Guide and Intervention

Circles and Circumference

Parts of Circles A **circle** consists of all points in a plane that are a given distance, called the **radius**, from a given point called the **center**.

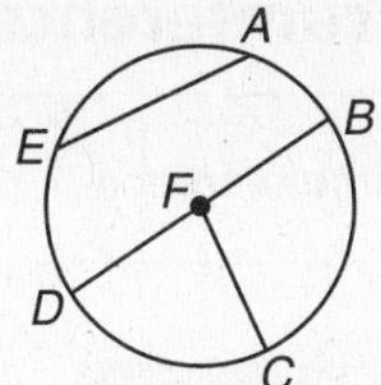

chord: $\overline{AE}$, $\overline{BD}$
radius: $\overline{FB}$, $\overline{FC}$, $\overline{FD}$
diameter: $\overline{BD}$

A segment or line can intersect a circle in several ways.

- A segment with endpoints that are the center of the circle and a point of the circle is a **radius**.
- A segment with endpoints that lie on the circle is a **chord**.
- A chord that contains the circle's center is a **diameter**.

Example

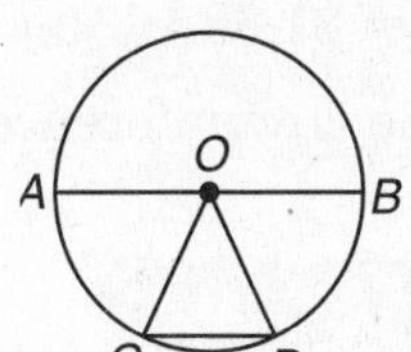

a. Name the circle.
The name of the circle is $\odot O$.

b. Name radii of the circle.
$\overline{AO}$, $\overline{BO}$, $\overline{CO}$, and $\overline{DO}$ are radii.

c. Name chords of the circle.
$\overline{AB}$ and $\overline{CD}$ are chords.

d. Name a diameter of the circle.
$\overline{AB}$ is a diameter.

Exercises

1. Name the circle.

2. Name radii of the circle.

3. Name chords of the circle.

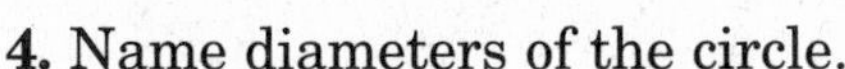

4. Name diameters of the circle.

5. Find AR if AB is 18 millimeters.

6. Find AR and AB if RY is 10 inches.

7. Is $\overline{AB} \cong \overline{XY}$? Explain.

Lesson 10-1

NAME ______________________ DATE ____________ PERIOD _____

10-1 Study Guide and Intervention *(continued)*

Circles and Circumference

Circumference The **circumference** of a circle is the distance around the circle.

Circumference	For a circumference of C units and a diameter of d units or a radius of r units, $C = \pi d$ or $C = 2\pi r$.

Example **Find the circumference of the circle to the nearest hundredth.**

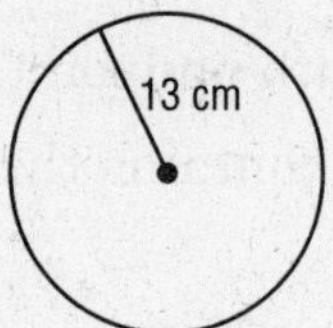

$C = 2\pi r$ Circumference formula

$= 2\pi(13)$ $r = 13$

≈ 81.68 Use a calculator.

The circumference is about 81.68 centimeters.

Exercises

Find the circumference of a circle with the given radius or diameter. Round to the nearest hundredth.

1. $r = 8$ cm

2. $r = 3\sqrt{2}$ ft

3. $r = 4.1$ cm

4. $d = 10$ in.

5. $d = \frac{1}{3}$ m

6. $d = 18$ yd

The radius, diameter, or circumference of a circle is given. Find the missing measures to the nearest hundredth.

7. $r = 4$ cm
$d =$ ____________, $C =$ ____________

8. $d = 6$ ft
$r =$ ____________, $C =$ ____________

9. $r = 12$ cm
$d =$ ____________, $C =$ ____________

10. $d = 15$ in.
$r =$ ____________, $C =$ ____________

Find the exact circumference of each circle.

11.

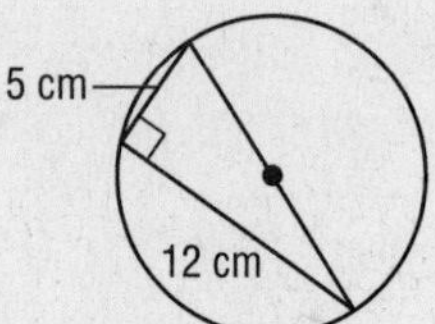

12.

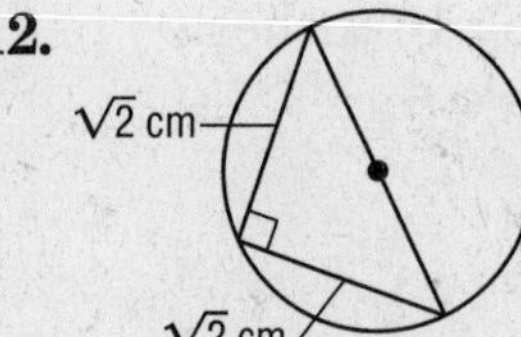

NAME ______________________ DATE __________ PERIOD _____

10-2 Study Guide and Intervention

Angles and Arcs

Angles and Arcs A **central angle** is an angle whose vertex is at the center of a circle and whose sides are radii. A central angle separates a circle into two arcs, a **major arc** and a **minor arc**.

$\widehat{GF}$ is a minor arc.
$\widehat{CHG}$ is a major arc.
$\angle GEF$ is a central angle.

Here are some properties of central angles and arcs.

- The sum of the measures of the central angles of a circle with no interior points in common is 360. — $m\angle HEC + m\angle CEF + m\angle FEG + m\angle GEH = 360$
- The measure of a minor arc equals the measure of its central angle. — $m\widehat{CF} = m\angle CEF$
- The measure of a major arc is 360 minus the measure of the minor arc. — $m\widehat{CGF} = 360 - m\widehat{CF}$
- Two arcs are congruent if and only if their corresponding central angles are congruent. — $\widehat{CF} \cong \widehat{FG}$ if and only if $\angle CEF \cong \angle FEG$.
- The measure of an arc formed by two adjacent arcs is the sum of the measures of the two arcs. **(Arc Addition Postulate)** — $m\widehat{CF} + m\widehat{FG} = m\widehat{CG}$

Example **In ⊙R, $m\angle ARB = 42$ and $\overline{AC}$ is a diameter. Find $m\widehat{AB}$ and $m\widehat{ACB}$.**

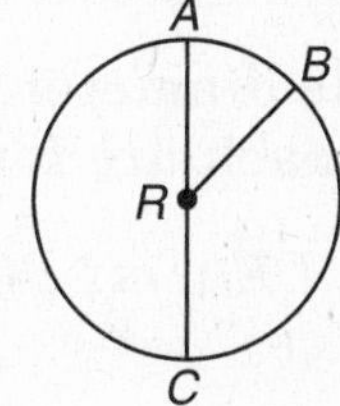

$\angle ARB$ is a central angle and $m\angle ARB = 42$, so $m\widehat{AB} = 42$.
Thus $m\widehat{ACB} = 360 - 42$ or 318.

Exercises

Find each measure.

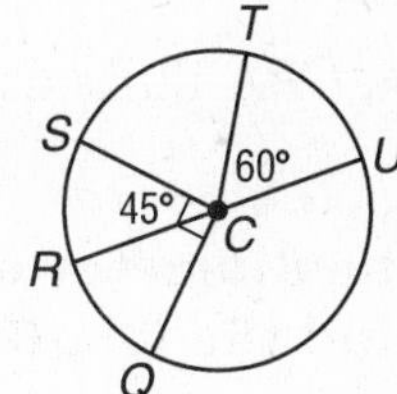

1. $m\angle SCT$ **2.** $m\angle SCU$

3. $m\angle SCQ$ **4.** $m\angle QCT$

If $m\angle BOA = 44$, find each measure.

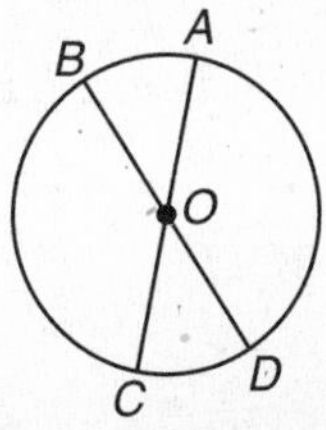

5. $m\widehat{BA}$ **6.** $m\widehat{BC}$

7. $m\widehat{CD}$ **8.** $m\widehat{ACB}$

9. $m\widehat{BCD}$ **10.** $m\widehat{AD}$

NAME ______________________ DATE __________ PERIOD _____

10-2 Study Guide and Intervention *(continued)*

Angles and Arcs

Arc Length An arc is part of a circle and its length is a part of the circumference of the circle.

Example **In $\odot R$, $m\angle ARB = 135$, $RB = 8$, and $\overline{AC}$ is a diameter. Find the length of $\widehat{AB}$.**

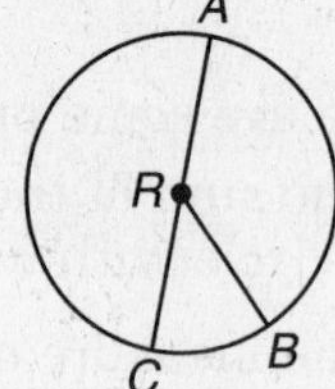

$m\angle ARB = 135$, so $m\widehat{AB} = 135$. Using the formula $C = 2\pi r$, the circumference is $2\pi(8)$ or 16π. To find the length of $\widehat{AB}$, write a proportion to compare each part to its whole.

$\frac{\text{length of } \widehat{AB}}{\text{circumference}} = \frac{\text{degree measure of arc}}{\text{degree measure of circle}}$ Proportion

$\frac{\ell}{16\pi} = \frac{135}{360}$ Substitution

$\ell = \frac{(16\pi)(135)}{360}$ Multiply each side by 16π.

$= 6\pi$ Simplify.

The length of $\widehat{AB}$ is 6π or about 18.85 units.

Exercises

The diameter of $\odot O$ is 24 units long. Find the length of each arc for the given angle measure.

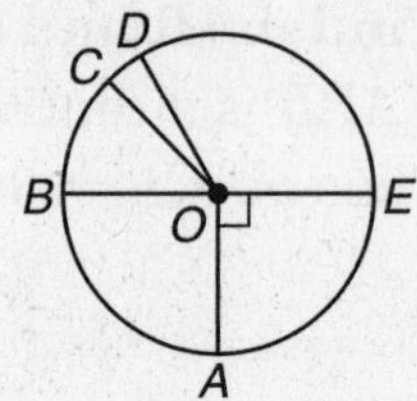

1. $\widehat{DE}$ if $m\angle DOE = 120$

2. $\widehat{DEA}$ if $m\angle DOE = 120$

3. $\widehat{BC}$ if $m\angle COB = 45$

4. $\widehat{CBA}$ if $m\angle COB = 45$

The diameter of $\odot P$ is 15 units long and $\angle SPT \cong \angle RPT$. Find the length of each arc for the given angle measure.

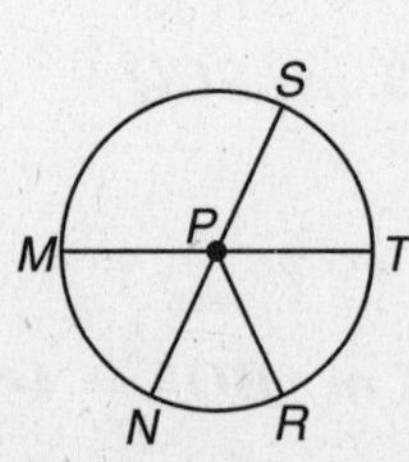

5. $\widehat{RT}$ if $m\angle SPT = 70$

6. $\widehat{NR}$ if $m\angle RPT = 50$

7. $\widehat{MST}$

8. $\widehat{MRS}$ if $m\angle MPS = 140$

10-3 Study Guide and Intervention

Arcs and Chords

Arcs and Chords Points on a circle determine both chords and arcs. Several properties are related to points on a circle.

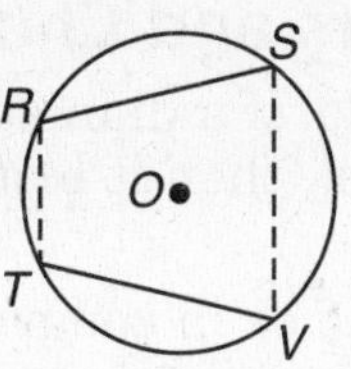

$\widehat{RS} \cong \widehat{TV}$ if and only if $\overline{RS} \cong \overline{TV}$.
RSVT is inscribed in ⊙*O*.
⊙*O* is circumscribed about *RSVT*.

- In a circle or in congruent circles, two minor arcs are congruent if and only if their corresponding chords are congruent.
- If all the vertices of a polygon lie on a circle, the polygon is said to be **inscribed** in the circle and the circle is **circumscribed** about the polygon.

Example **Trapezoid *ABCD* is inscribed in ⊙*O*. If $\overline{AB} \cong \overline{BC} \cong \overline{CD}$ and $m\widehat{BC} = 50$, what is $m\widehat{APD}$?**

Chords $\overline{AB}$, $\overline{BC}$, and $\overline{CD}$ are congruent, so $\widehat{AB}$, $\widehat{BC}$, and $\widehat{CD}$ are congruent. $m\widehat{BC} = 50$, so $m\widehat{AB} + m\widehat{BC} + m\widehat{CD} = 50 + 50 + 50 = 150$. Then $m\widehat{APD} = 360 - 150$ or 210.

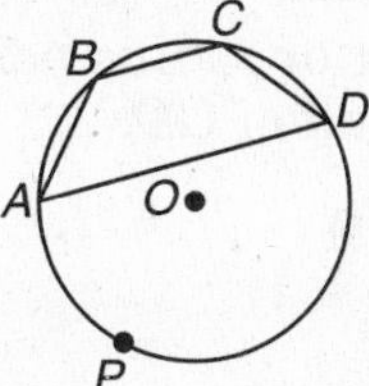

Exercises

Each regular polygon is inscribed in a circle. Determine the measure of each arc that corresponds to a side of the polygon.

1. hexagon

2. pentagon

3. triangle

4. square

5. octagon

6. 36-gon

Determine the measure of each arc of the circle circumscribed about the polygon.

7.

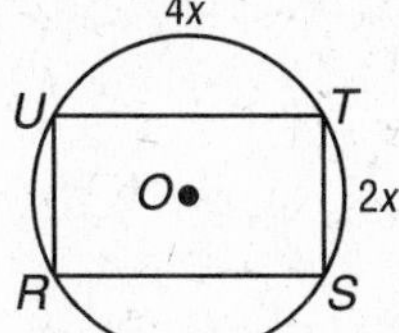

8.

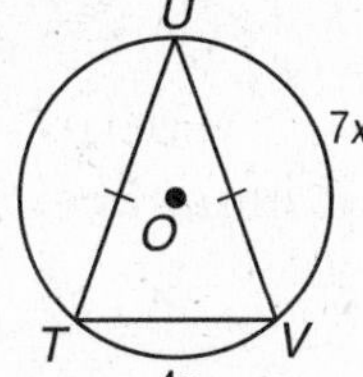

9.

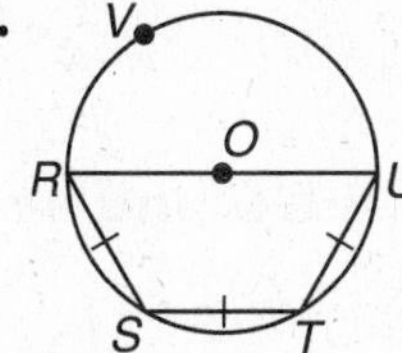

Lesson 10-3

10-3 Study Guide and Intervention *(continued)*

Arcs and Chords

Diameters and Chords

- In a circle, if a diameter is perpendicular to a chord, then it bisects the chord and its arc.
- In a circle or in congruent circles, two chords are congruent if and only if they are equidistant from the center.

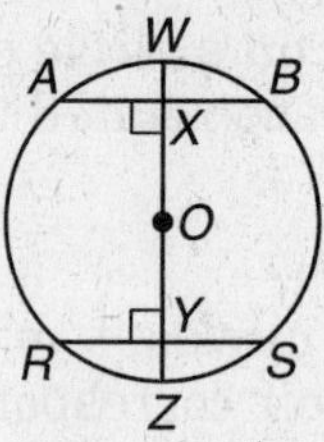

If $\overline{WZ} \perp \overline{AB}$, then $\overline{AX} \cong \overline{XB}$ and $\widehat{AW} \cong \widehat{WB}$.
If $OX = OY$, then $\overline{AB} \cong \overline{RS}$.
If $\overline{AB} \cong \overline{RS}$, then $\overline{AB}$ and $\overline{RS}$ are equidistant from point O.

Example **In $\odot O$, $\overline{CD} \perp \overline{OE}$, $OD = 15$, and $CD = 24$. Find x.**

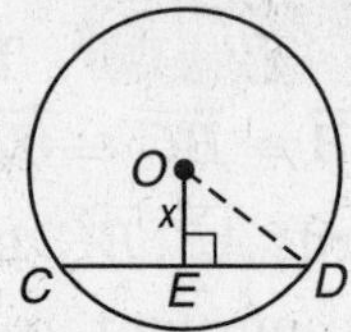

A diameter or radius perpendicular to a chord bisects the chord, so ED is half of CD.

$$ED = \frac{1}{2}(24)$$
$$= 12$$

Use the Pythagorean Theorem to find x in $\triangle OED$.

$(OE)^2 + (ED)^2 = (OD)^2$	Pythagorean Theorem
$x^2 + 12^2 = 15^2$	Substitution
$x^2 + 144 = 225$	Multiply.
$x^2 = 81$	Subtract 144 from each side.
$x = 9$	Take the square root of each side.

Exercises

In $\odot P$, $CD = 24$ and $m\widehat{CY} = 45$. Find each measure.

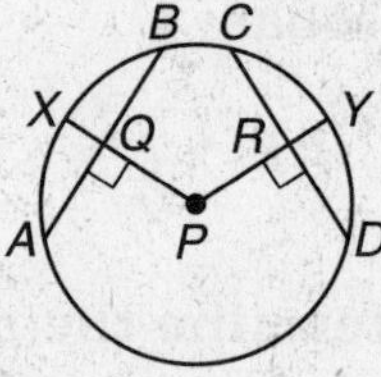

1. AQ **2.** RC **3.** QB

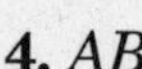

4. AB **5.** $m\widehat{DY}$ **6.** $m\widehat{AB}$

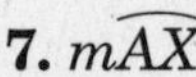

7. $m\widehat{AX}$ **8.** $m\widehat{XB}$ **9.** $m\widehat{CD}$

In $\odot G$, $DG = GU$ and $AC = RT$. Find each measure.

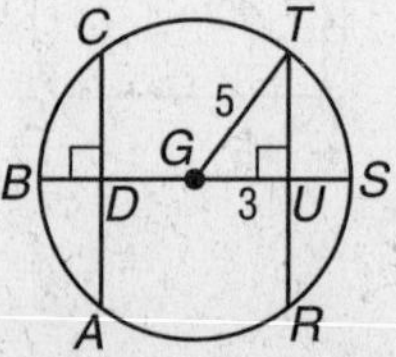

10. TU **11.** TR **12.** $m\widehat{TS}$

13. CD **14.** GD **15.** $m\widehat{AB}$

16. A chord of a circle 20 inches long is 24 inches from the center of a circle. Find the length of the radius.

10-4 Study Guide and Intervention

Inscribed Angles

Inscribed Angles An **inscribed angle** is an angle whose vertex is on a circle and whose sides contain chords of the circle. In $\odot G$, inscribed $\angle DEF$ **intercepts** $\widehat{DF}$.

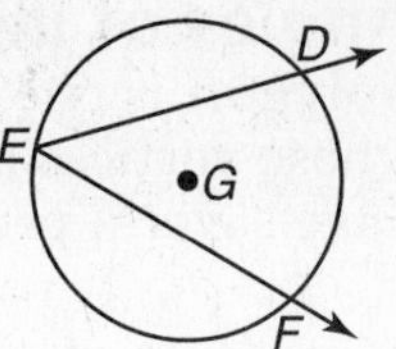

$m\angle DEF = \frac{1}{2}m\widehat{DF}$

Inscribed Angle Theorem	If an angle is inscribed in a circle, then the measure of the angle equals one-half the measure of its intercepted arc.

Example **In $\odot G$ above, $m\widehat{DF} = 90$. Find $m\angle DEF$.**

$\angle DEF$ is an inscribed angle so its measure is half of the intercepted arc.

$$m\angle DEF = \frac{1}{2}m\widehat{DF}$$
$$= \frac{1}{2}(90) \text{ or } 45$$

Exercises

Use $\odot P$ for Exercises 1–10. In $\odot P$, $\overline{RS} \parallel \overline{TV}$ and $\overline{RT} \cong \overline{SV}$.

1. Name the intercepted arc for $\angle RTS$.

2. Name an inscribed angle that intercepts $\widehat{SV}$.

In $\odot P$, $m\widehat{SV} = 120$ and $m\angle RPS = 76$. Find each measure.

3. $m\angle PRS$

4. $m\widehat{RSV}$

5. $m\widehat{RT}$

6. $m\angle RVT$

7. $m\angle QRS$

8. $m\angle STV$

9. $m\widehat{TV}$

10. $m\angle SVT$

Lesson 10-4

10-4 Study Guide and Intervention *(continued)*

Inscribed Angles

Angles of Inscribed Polygons An **inscribed polygon** is one whose sides are chords of a circle and whose vertices are points on the circle. Inscribed polygons have several properties.

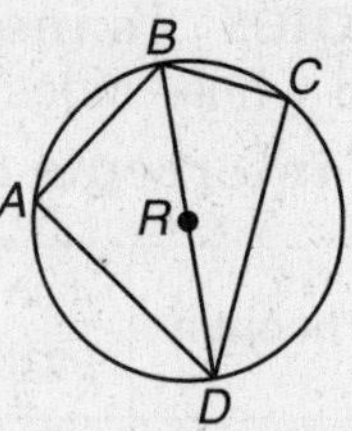

- If an angle of an inscribed polygon intercepts a semicircle, the angle is a right angle.
- If a quadrilateral is inscribed in a circle, then its opposite angles are supplementary.

If $\overset{\frown}{BCD}$ is a semicircle, then $m\angle BCD = 90$.

For inscribed quadrilateral $ABCD$, $m\angle A + m\angle C = 180$ and $m\angle ABC + m\angle ADC = 180$.

Example **In ⊙R above, $BC = 3$ and $BD = 5$. Find each measure.**

a. $m\angle C$

$\angle C$ intercepts a semicircle. Therefore $\angle C$ is a right angle and $m\angle C = 90$.

b. CD

$\triangle BCD$ is a right triangle, so use the Pythagorean Theorem to find CD.

$$(CD)^2 + (BC)^2 = (BD)^2$$
$$(CD)^2 + 3^2 = 5^2$$
$$(CD)^2 = 25 - 9$$
$$(CD)^2 = 16$$
$$CD = 4$$

Exercises

Find the measure of each angle or segment for each figure.

1. $m\angle X$, $m\angle Y$

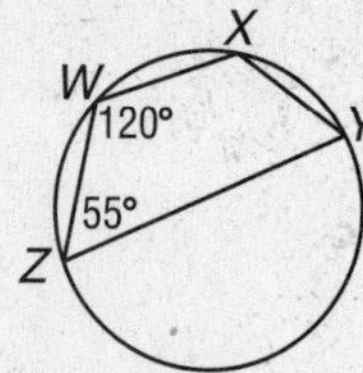

2. AD

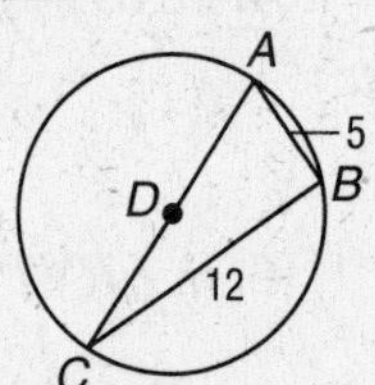

3. $m\angle 1$, $m\angle 2$

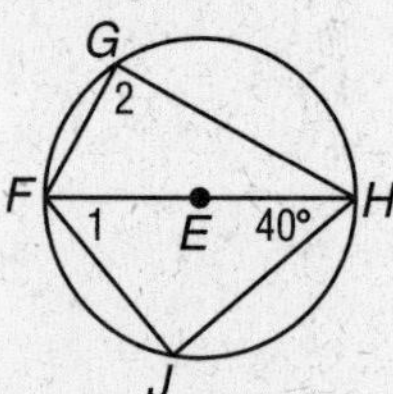

4. $m\angle 1$, $m\angle 2$

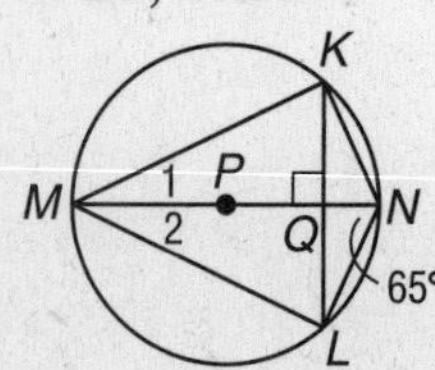

5. AB, AC

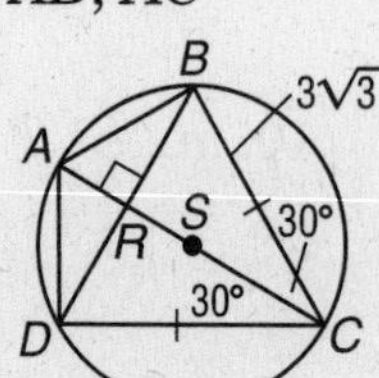

6. $m\angle 1$, $m\angle 2$

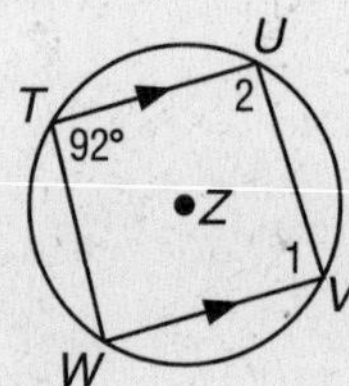

NAME ______________________ DATE __________ PERIOD _____

10-5 Study Guide and Intervention

Tangents

Tangents A tangent to a circle intersects the circle in exactly one point, called the **point of tangency**. There are three important relationships involving tangents.

- If a line is tangent to a circle, then it is perpendicular to the radius drawn to the point of tangency.
- If a line is perpendicular to a radius of a circle at its endpoint on the circle, then the line is a tangent to the circle.
- If two segments from the same exterior point are tangent to a circle, then they are congruent.

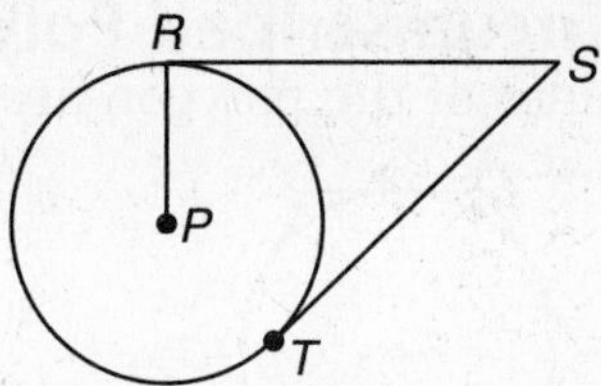

$\overline{RP} \perp \overline{SR}$ if and only if $\overline{SR}$ is tangent to $\odot P$.

If $\overline{SR}$ and $\overline{ST}$ are tangent to $\odot P$, then $\overline{SR} \cong \overline{ST}$.

Example $\overline{AB}$ **is tangent to** $\odot C$**. Find** x**.**

$\overline{AB}$ is tangent to $\odot C$, so $\overline{AB}$ is perpendicular to radius $\overline{BC}$. $\overline{CD}$ is a radius, so $CD = 8$ and $AC = 9 + 8$ or 17. Use the Pythagorean Theorem with right $\triangle ABC$.

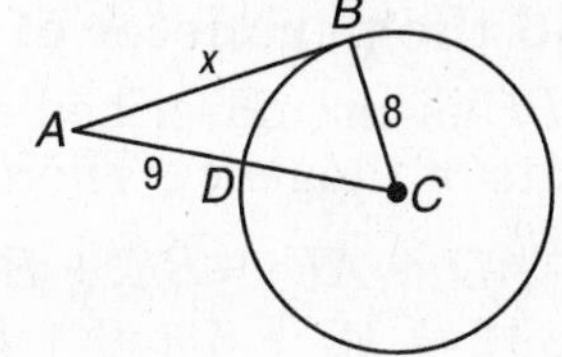

$(AB)^2 + (BC)^2 = (AC)^2$	Pythagorean Theorem
$x^2 + 8^2 = 17^2$	Substitution
$x^2 + 64 = 289$	Multiply.
$x^2 = 225$	Subtract 64 from each side.
$x = 15$	Take the square root of each side.

Exercises

Find x**. Assume that segments that appear to be tangent are tangent.**

1.

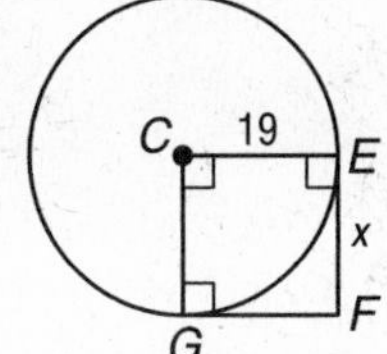

2.

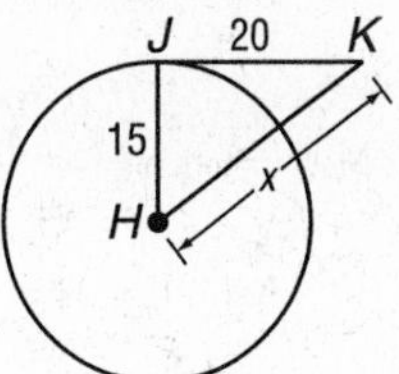

3.

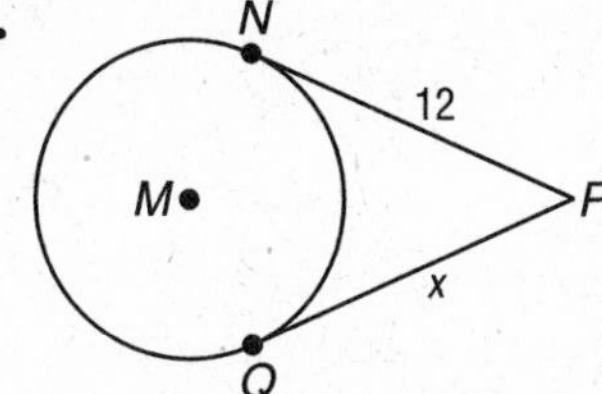

4.

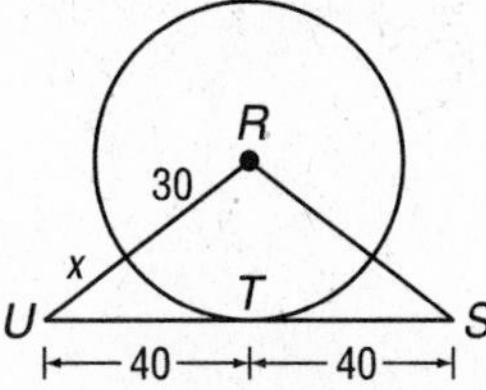

5.

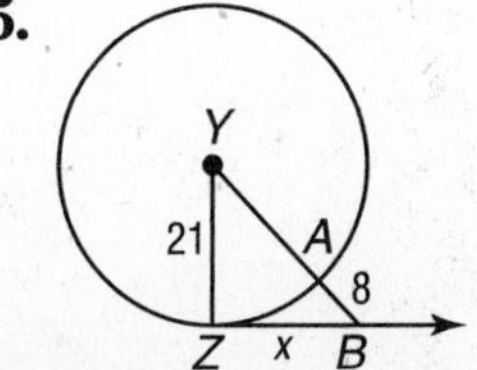

6.

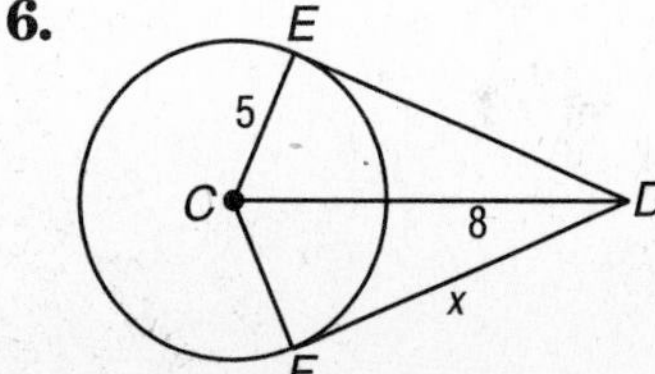

Lesson 10-5

10-5 Study Guide and Intervention *(continued)*

Tangents

Circumscribed Polygons When a polygon is circumscribed about a circle, all of the sides of the polygon are tangent to the circle.

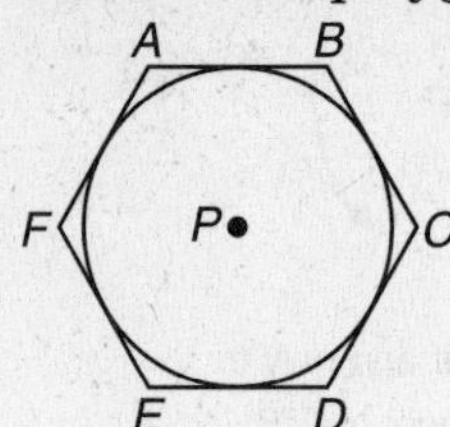

Hexagon *ABCDEF* is circumscribed about ⊙*P*. $\overline{AB}$, $\overline{BC}$, $\overline{CD}$, $\overline{DE}$, $\overline{EF}$, and $\overline{FA}$ are tangent to ⊙*P*.

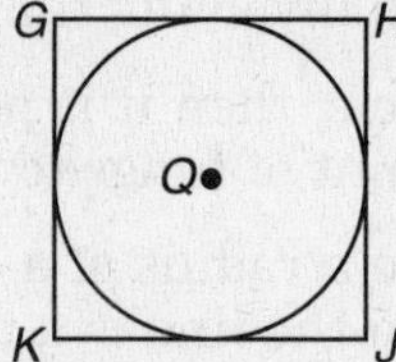

Square *GHJK* is circumscribed about ⊙*Q*. $\overline{GH}$, $\overline{JH}$, $\overline{JK}$, and $\overline{KG}$ are tangent to ⊙*Q*.

Example **△*ABC* is circumscribed about ⊙*O*. Find the perimeter of △*ABC*.**

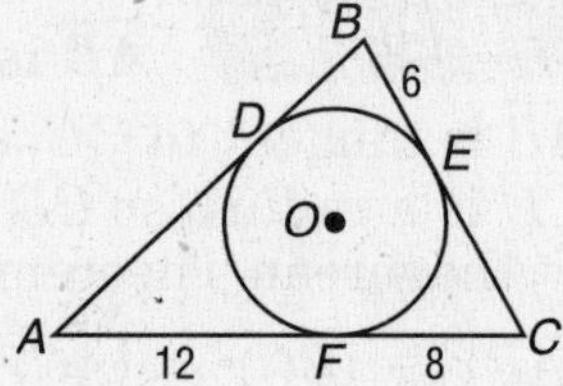

△*ABC* is circumscribed about ⊙*O*, so points *D*, *E*, and *F* are points of tangency. Therefore $AD = AF$, $BE = BD$, and $CF = CE$.

$P = AD + AF + BE + BD + CF + CE$
$= 12 + 12 + 6 + 6 + 8 + 8$
$= 52$

The perimeter is 52 units.

Exercises

Find *x*. Assume that segments that appear to be tangent are tangent.

1.

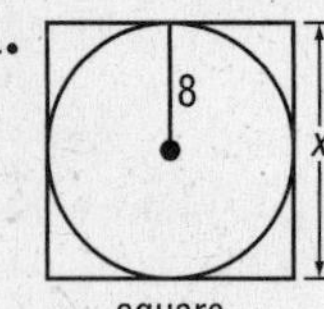

2.

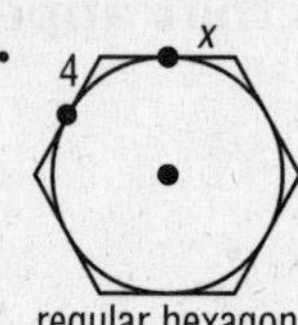

3.

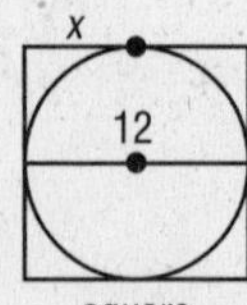

4.

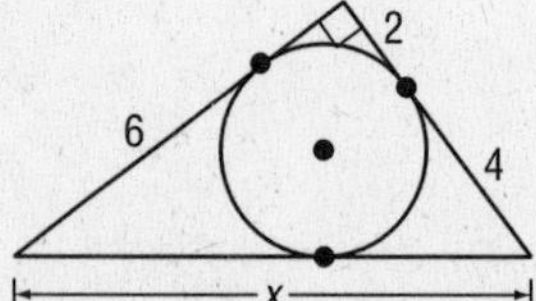

5.

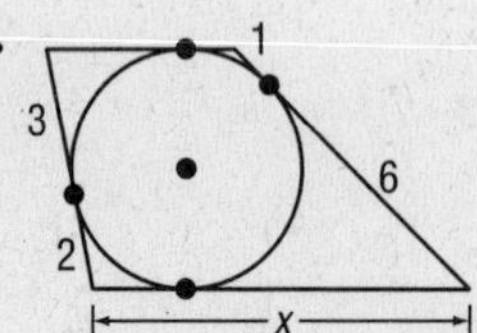

6.

NAME ______________________ DATE __________ PERIOD _____

10-6 Study Guide and Intervention

Secants, Tangents, and Angle Measures

Lesson 10-6

Intersections On or Inside a Circle A line that intersects a circle in exactly two points is called a **secant**. The measures of angles formed by secants and tangents are related to intercepted arcs.

- If two secants intersect in the interior of a circle, then the measure of the angle formed is one-half the sum of the measure of the arcs intercepted by the angle and its vertical angle.

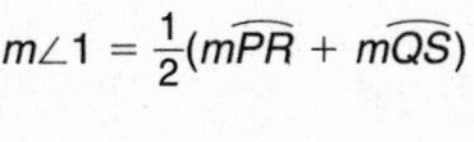

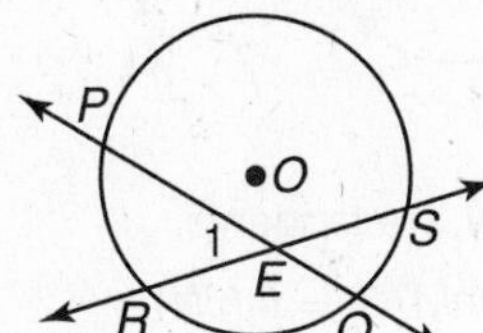

- If a secant and a tangent intersect at the point of tangency, then the measure of each angle formed is one-half the measure of its intercepted arc.

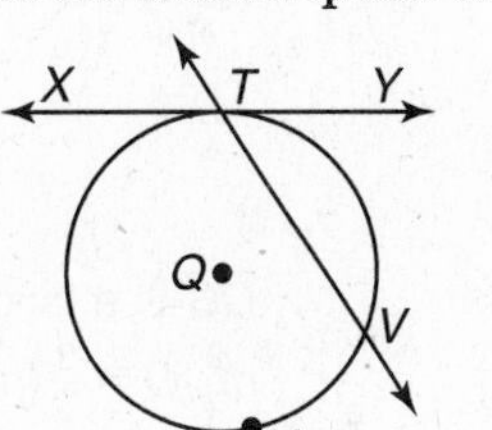

$m\angle XTV = \frac{1}{2}m\overset{\frown}{TUV}$

$m\angle YTV = \frac{1}{2}m\overset{\frown}{TV}$

Example 1 **Find *x*.**

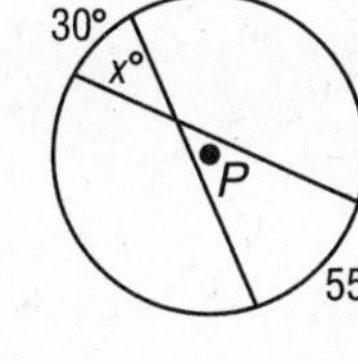

The two secants intersect inside the circle, so x is equal to one-half the sum of the measures of the arcs intercepted by the angle and its vertical angle.

$x = \frac{1}{2}(30 + 55)$

$= \frac{1}{2}(85)$

$= 42.5$

Example 2 **Find *y*.**

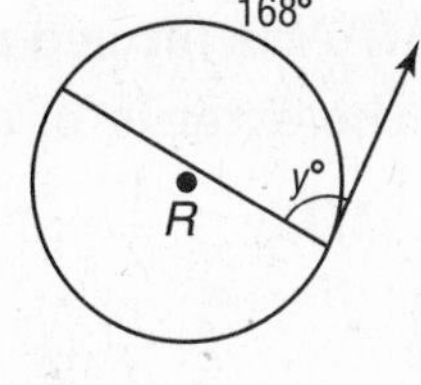

The secant and the tangent intersect at the point of tangency, so the measure of the angle is one-half the measure of its intercepted arc.

$y = \frac{1}{2}(168)$

$= 84$

Exercises

Find each measure.

1. $m\angle 1$

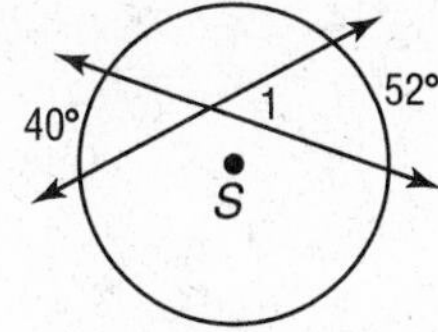

2. $m\angle 2$

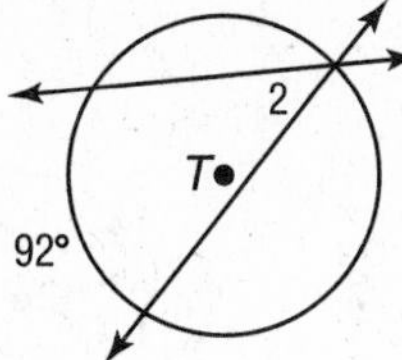

3. $m\angle 3$

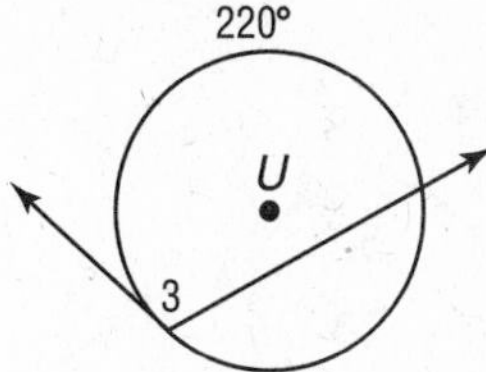

4. $m\angle 4$

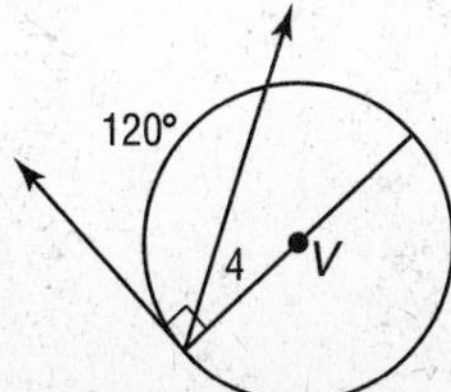

5. $m\angle 5$

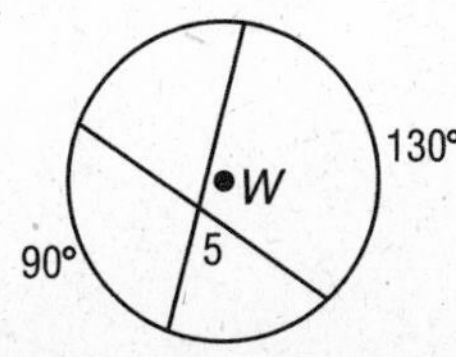

6. $m\angle 6$

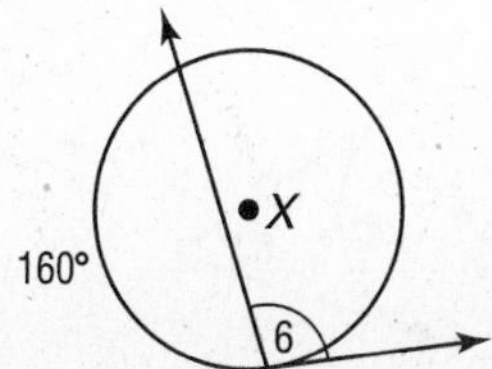

10-6 Study Guide and Intervention *(continued)*

Secants, Tangents, and Angle Measures

Intersections Outside a Circle If secants and tangents intersect outside a circle, they form an angle whose measure is related to the intercepted arcs.

If two secants, a secant and a tangent, or two tangents intersect in the exterior of a circle, then the measure of the angle formed is one-half the positive difference of the measures of the intercepted arcs.

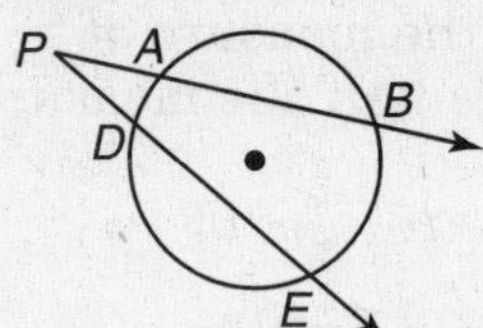

$\overrightarrow{PB}$ and $\overrightarrow{PE}$ are secants.
$m\angle P = \frac{1}{2}(m\widehat{BE} - m\widehat{AD})$

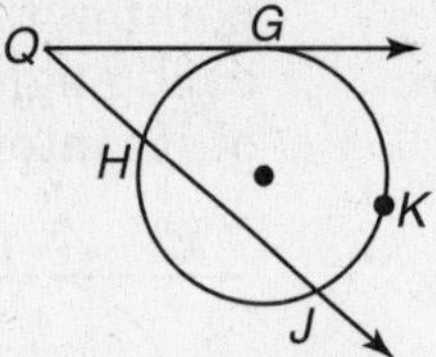

$\overrightarrow{QG}$ is a tangent. $\overrightarrow{QJ}$ is a secant.
$m\angle Q = \frac{1}{2}(m\widehat{GKJ} - m\widehat{GH})$

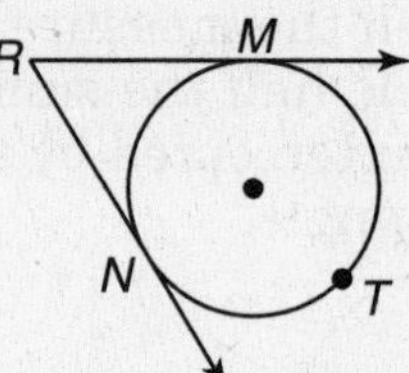

$\overrightarrow{RM}$ and $\overrightarrow{RN}$ are tangents.
$m\angle R = \frac{1}{2}(m\widehat{MTN} - m\widehat{MN})$

Example **Find $m\angle MPN$.**

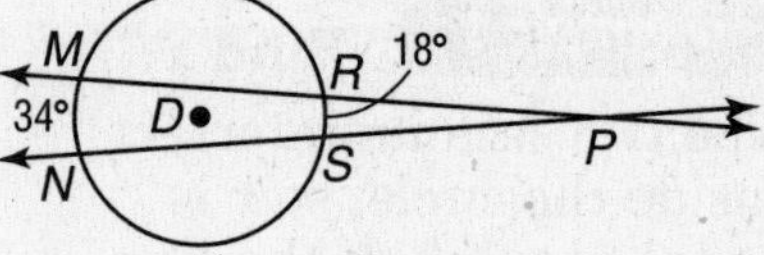

$\angle MPN$ is formed by two secants that intersect in the exterior of a circle.

$$m\angle MPN = \frac{1}{2}(m\widehat{MN} - m\widehat{RS})$$
$$= \frac{1}{2}(34 - 18)$$
$$= \frac{1}{2}(16) \text{ or } 8$$

The measure of the angle is 8.

Exercises

Find each measure.

1. $m\angle 1$

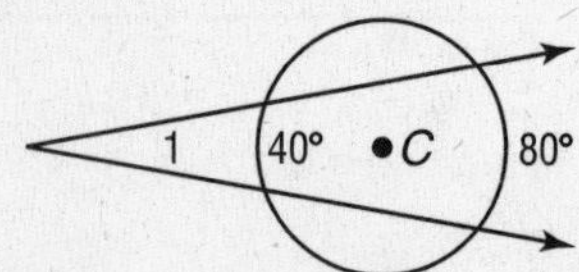

2. $m\angle 2$

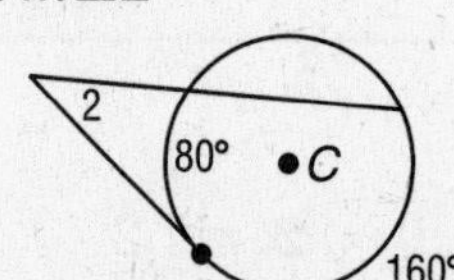

3. $m\angle 3$

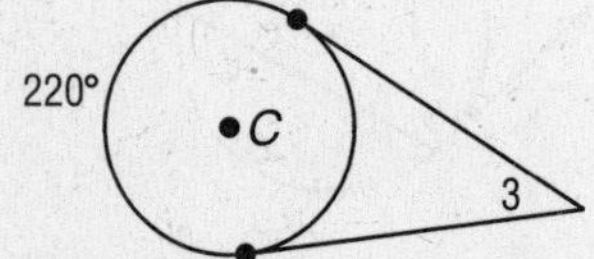

4. x

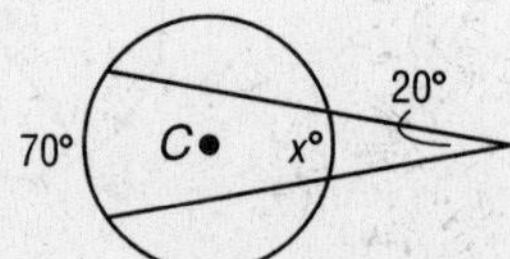

5. x

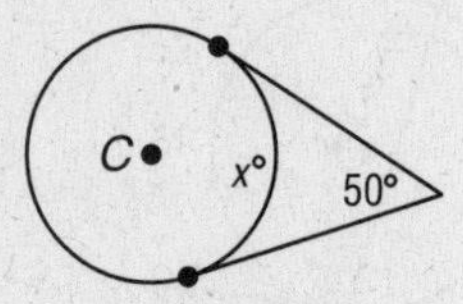

6. x

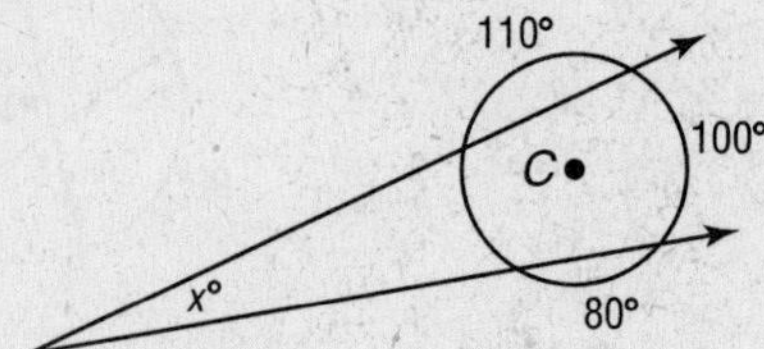

10-7 Study Guide and Intervention

Special Segments in a Circle

Segments Intersecting Inside a Circle If two chords intersect in a circle, then the products of the measures of the chords are equal.

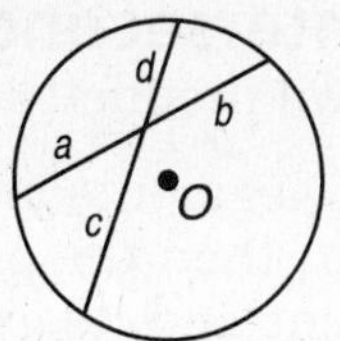

$a \cdot b = c \cdot d$

Example **Find x.**

The two chords intersect inside the circle, so the products $AB \cdot BC$ and $EB \cdot BD$ are equal.

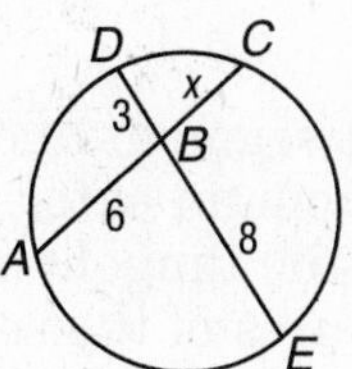

$AB \cdot BC = EB \cdot BD$

$AB \cdot BC = EB \cdot BD$

$6 \cdot x = 8 \cdot 3$ Substitution

$6x = 24$ Simplify.

$x = 4$ Divide each side by 6.

Exercises

Find x to the nearest tenth.

1. 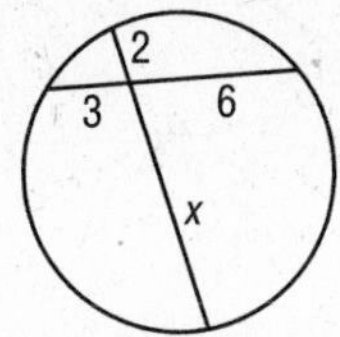

2. 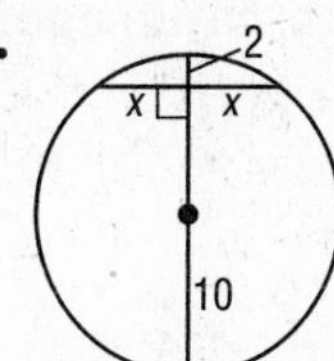

3. 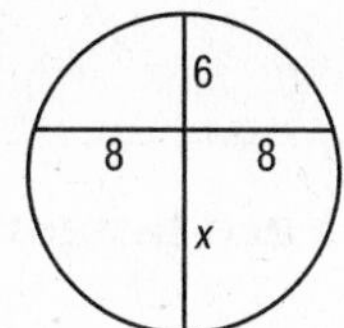

4. 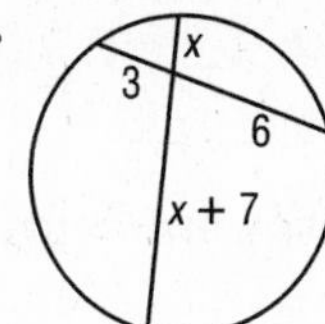

5. 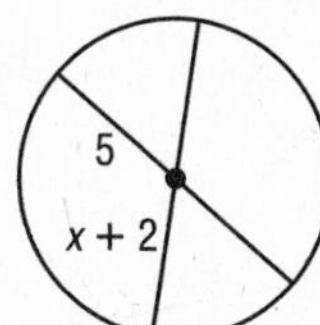

6. 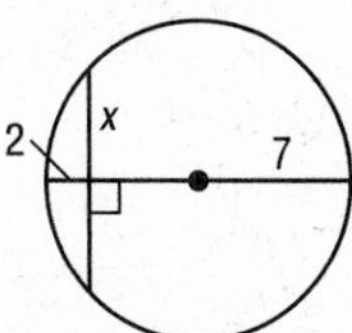

7. 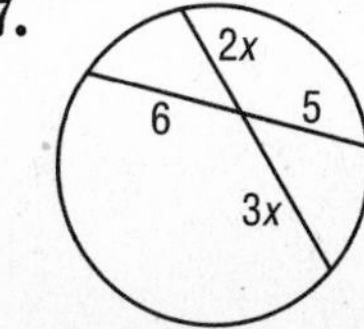

8. 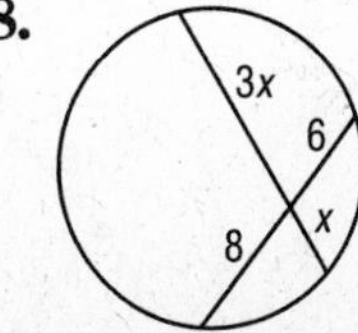

10-7 Study Guide and Intervention *(continued)*

Special Segments in a Circle

Segments Intersecting Outside a Circle If secants and tangents intersect outside a circle, then two products are equal.

- If two secant segments are drawn to a circle from an exterior point, then the product of the measures of one secant segment and its external secant segment is equal to the product of the measures of the other secant segment and its external secant segment.

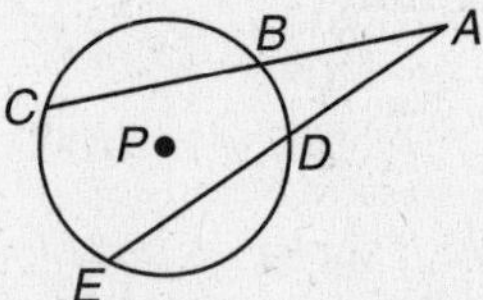

$\overline{AC}$ and $\overline{AE}$ are secant segments.
$\overline{AB}$ and $\overline{AD}$ are external secant segments.
$AC \cdot AB = AE \cdot AD$

- If a tangent segment and a secant segment are drawn to a circle from an exterior point, then the square of the measure of the tangent segment is equal to the product of the measures of the secant segment and its external secant segment.

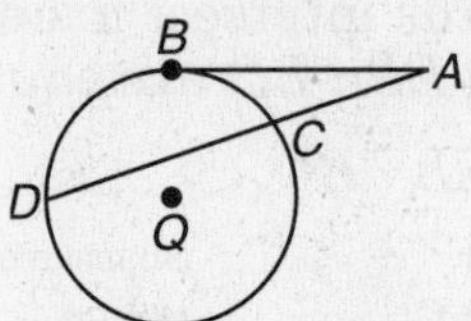

$\overline{AB}$ is a tangent segment.
$\overline{AD}$ is a secant segment.
$\overline{AC}$ is an external secant segment.
$(AB)^2 = AD \cdot AC$

Example **Find x to the nearest tenth.**

The tangent segment is $\overline{AB}$, the secant segment is $\overline{BD}$, and the external secant segment is $\overline{BC}$.

$(AB)^2 = BC \cdot BD$
$(18)^2 = 15(15 + x)$
$324 = 225 + 15x$
$99 = 15x$
$6.6 = x$

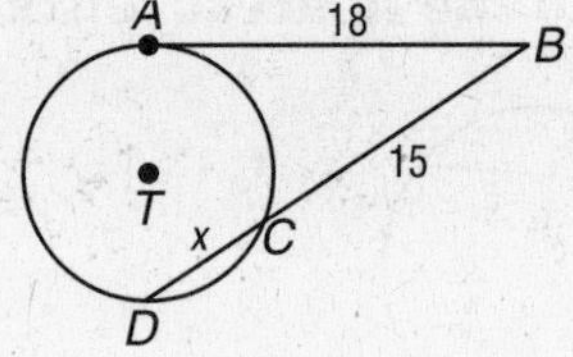

Exercises

Find x to the nearest tenth. Assume segments that appear to be tangent are tangent.

1.
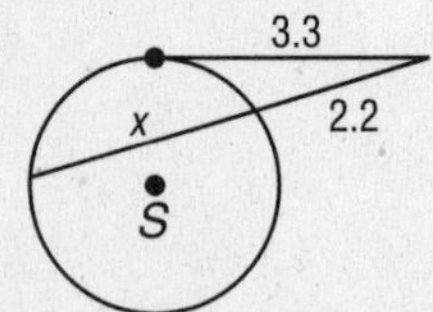

2.
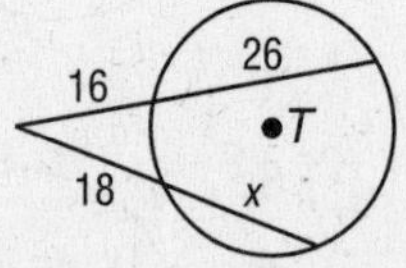

3.
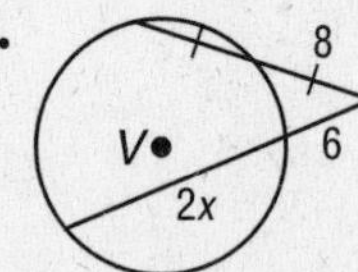

4.
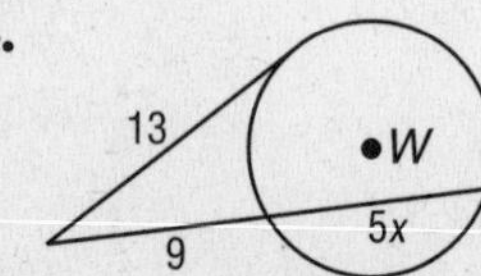

5.
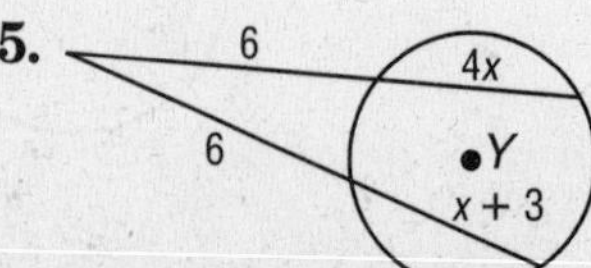

6.
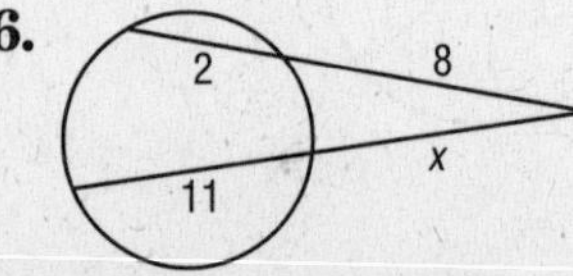

7.
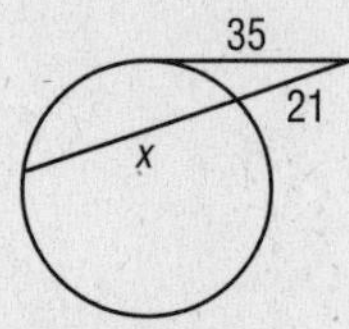

8.
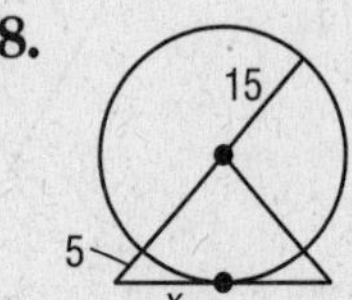

9.
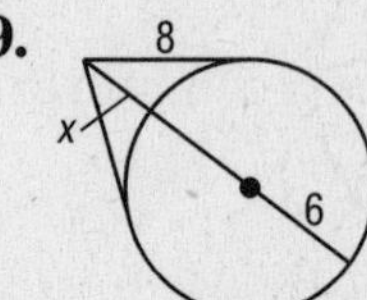

NAME ______________________ DATE ____________ PERIOD _____

10-8 Study Guide and Intervention

Equations of Circles

Equation of a Circle A **circle** is the locus of points in a plane equidistant from a given point. You can use this definition to write an equation of a circle.

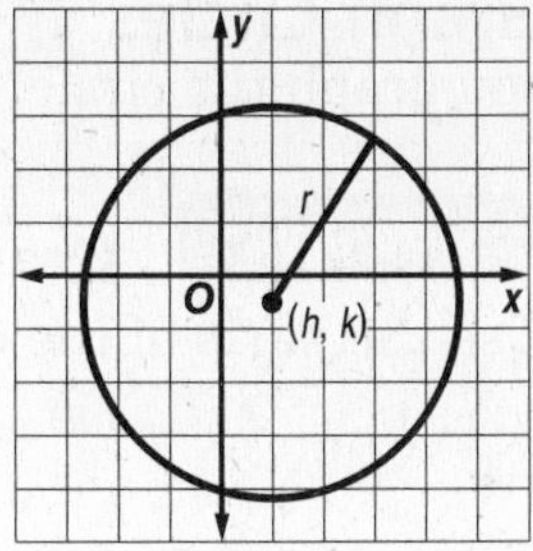

Standard Equation of a Circle	An equation for a circle with center at (h, k) and a radius of r units is $(x - h)^2 + (y - k)^2 = r^2$.

Example **Write an equation for a circle with center (−1, 3) and radius 6.**

Use the formula $(x - h)^2 + (y - k)^2 = r^2$ with $h = -1$, $k = 3$, and $r = 6$.

$(x - h)^2 + (y - k)^2 = r^2$ Equation of a circle

$(x - (-1))^2 + (y - 3)^2 = 6^2$ Substitution

$(x + 1)^2 + (y - 3)^2 = 36$ Simplify.

Exercises

Write an equation for each circle.

1. center at $(0, 0)$, $r = 8$

2. center at $(-2, 3)$, $r = 5$

3. center at $(2, -4)$, $r = 1$

4. center at $(-1, -4)$, $r = 2$

5. center at $(-2, -6)$, diameter $= 8$

6. center at $\left(-\frac{1}{2}, \frac{1}{4}\right)$, $r = \sqrt{3}$

7. center at the origin, diameter $= 4$

8. center at $\left(1, -\frac{5}{8}\right)$, $r = \sqrt{5}$

9. Find the center and radius of a circle with equation $x^2 + y^2 = 20$.

10. Find the center and radius of a circle with equation $(x + 4)^2 + (y + 3)^2 = 16$.

NAME ______________________________ DATE ____________ PERIOD _____

10-8 Study Guide and Intervention *(continued)*

Equations of Circles

Graph Circles If you are given an equation of a circle, you can find information to help you graph the circle.

Example **Graph $(x + 3)^2 + (y - 1)^2 = 9$.**

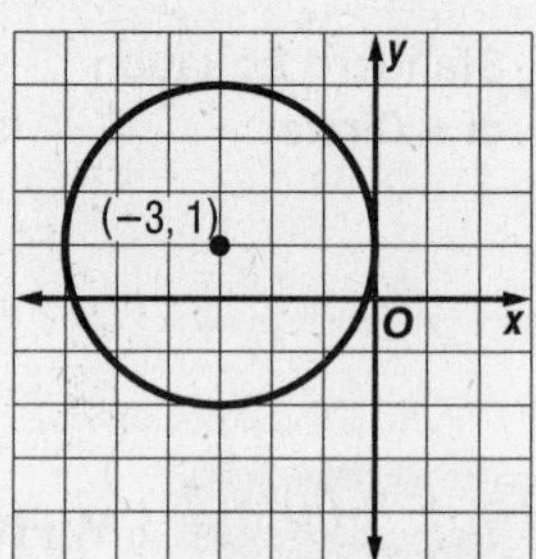

Use the parts of the equation to find (h, k) and r.

$(x - h)^2 + (y - k)^2 = r^2$

$(x - h)^2 = (x + 3)^2$	$(y - k)^2 = (y - 1)^2$	$r^2 = 9$
$x - h = x + 3$	$y - k = y - 1$	$r = 3$
$-h = 3$	$-k = -1$	
$h = -3$	$k = 1$	

The center is at $(-3, 1)$ and the radius is 3. Graph the center. Use a compass set at a radius of 3 grid squares to draw the circle.

Exercises

Graph each equation.

1. $x^2 + y^2 = 16$

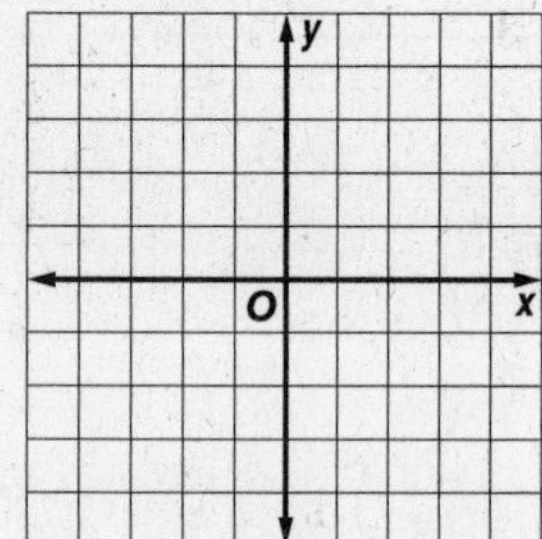

2. $(x - 2)^2 + (y - 1)^2 = 9$

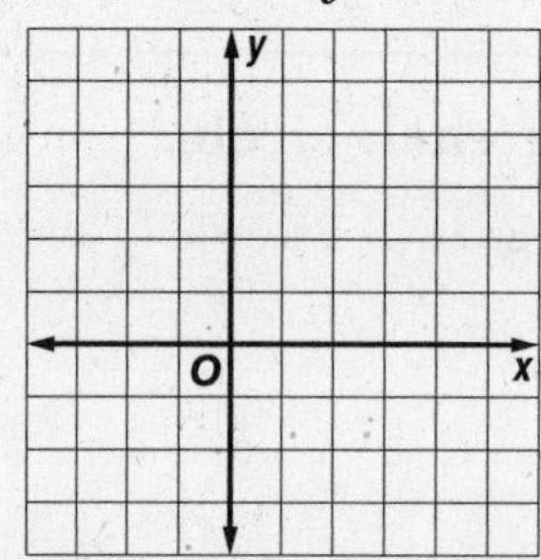

3. $(x + 2)^2 + y^2 = 16$

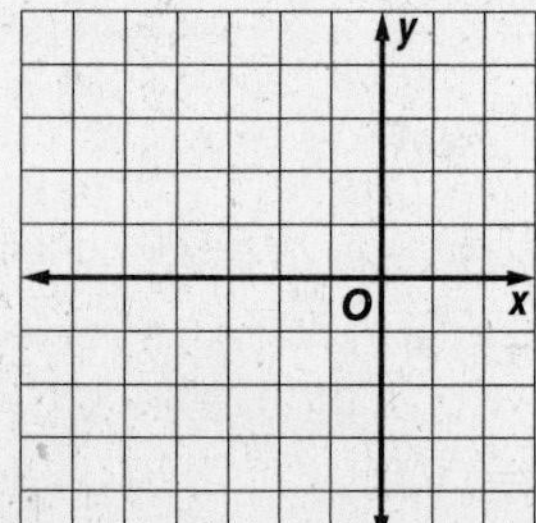

4. $(x + 1)^2 + (y - 2)^2 = 6.25$

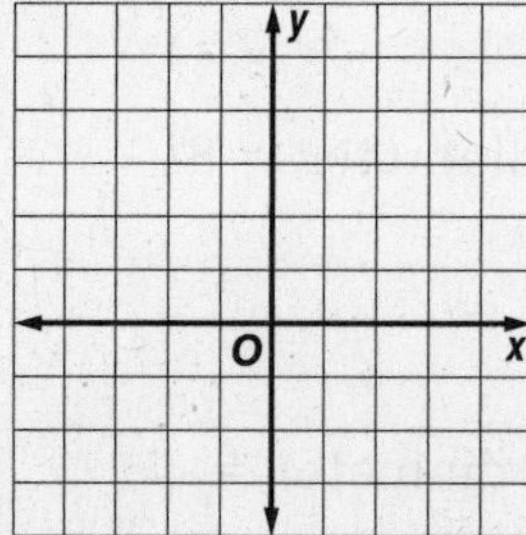

5. $\left(x + \frac{1}{2}\right)^2 + \left(y - \frac{1}{4}\right)^2 = 4$

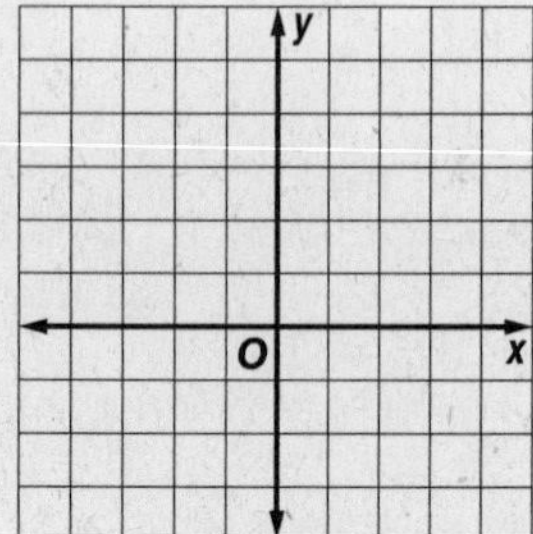

6. $x^2 + (y - 1)^2 = 9$

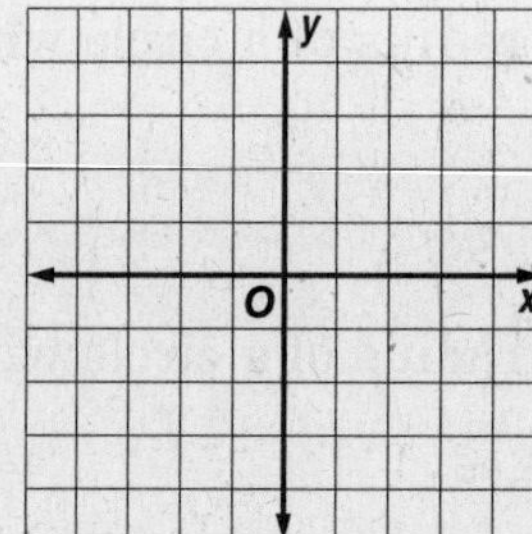

11-1 Study Guide and Intervention

Areas of Parallelograms

Areas of Parallelograms A parallelogram is a quadrilateral with both pairs of opposite sides parallel. Any side of a parallelogram can be called a **base**. Each base has a corresponding **altitude**, and the length of the altitude is the **height** of the parallelogram. The area of a parallelogram is the product of the base and the height.

Area of a Parallelogram	If a parallelogram has an area of A square units, a base of b units, and a height of h units, then $A = bh$.

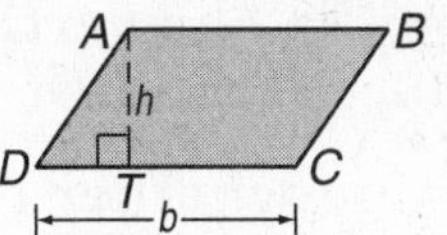

The area of parallelogram $ABCD$ is $CD \cdot AT$.

Example **Find the area of parallelogram *EFGH*.**

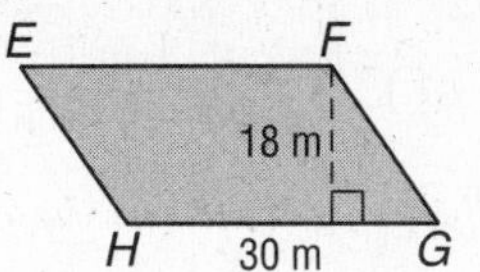

$A = bh$ Area of a parallelogram

$= 30(18)$ $b = 30$, $h = 18$

$= 540$ Multiply.

The area is 540 square meters.

Exercises

Find the area of each parallelogram.

1.

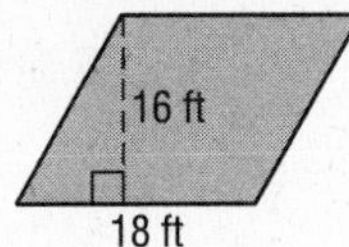

2.

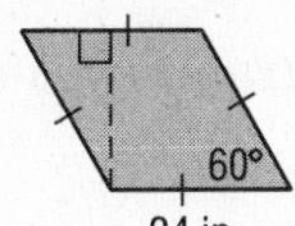

3.

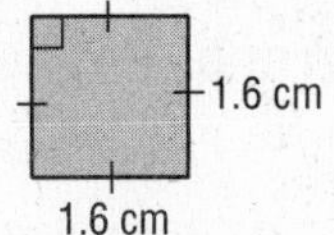

Find the area of each shaded region.

4. *WXYZ* and *ABCD* are rectangles.

W 32 cm X

A B

16 cm 5 cm

D 12 cm C

Z Y

5. All angles are right angles.

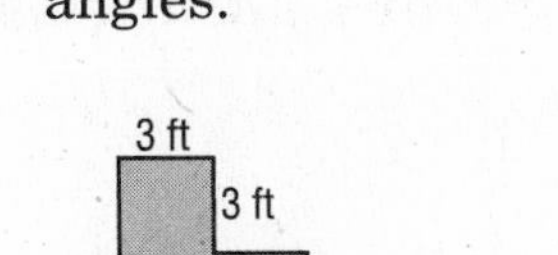

6. *EFGH* and *NOPQ* are rectangles; *JKLM* is a square.

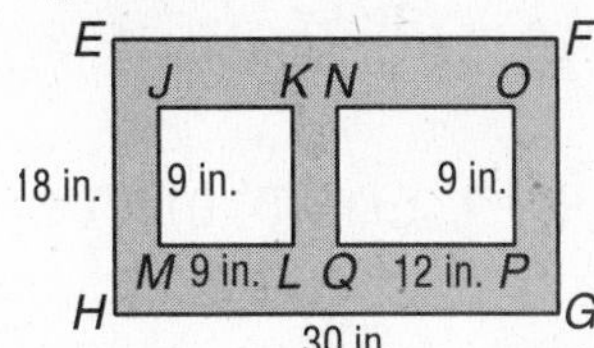

7. The area of a parallelogram is 3.36 square feet. The base is 2.8 feet. If the measures of the base and height are each doubled, find the area of the resulting parallelogram.

8. A rectangle is 4 meters longer than it is wide. The area of the rectangle is 252 square meters. Find the length.

11-1 Study Guide and Intervention *(continued)*

Areas of Parallelograms

Parallelograms on the Coordinate Plane To find the area of a quadrilateral on the coordinate plane, use the Slope Formula, the Distance Formula, and properties of parallelograms, rectangles, squares, and rhombi.

Example **The vertices of a quadrilateral are $A(-2, 2)$, $B(4, 2)$, $C(5, -1)$, and $D(-1, -1)$.**

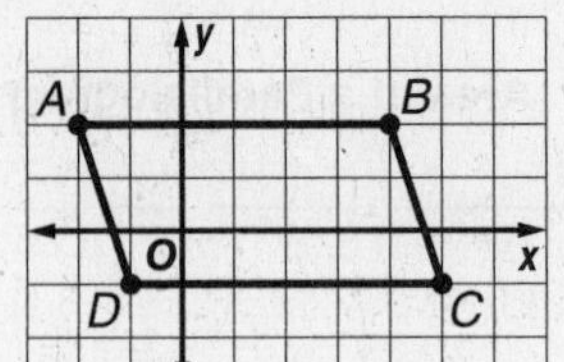

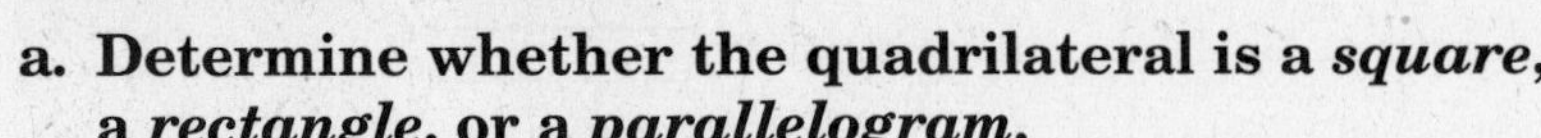

a. Determine whether the quadrilateral is a *square*, a *rectangle*, or a *parallelogram*.

Graph the quadrilateral. Then determine the slope of each side.

slope of $\overline{AB} = \frac{2 - 2}{4 - (-2)}$ or 0

slope of $\overline{CD} = \frac{-1 - (-1)}{-1 - 5}$ or 0

slope $\overline{AD} = \frac{2 - (-1)}{-2 - (-1)}$ or -3

slope $\overline{BC} = \frac{-1 - 2}{5 - 4}$ or -3

Opposite sides have the same slope. The slopes of consecutive sides are not negative reciprocals of each other, so consecutive sides are not perpendicular. *ABCD* is a parallelogram; it is not a rectangle or a square.

b. Find the area of *ABCD*.

From the graph, the height of the parallelogram is 3 units and $AB = |4 - (-2)| = 6$.

$A = bh$	Area of a parallelogram
$= 6(3)$	$b = 6$, $h = 3$
$= 18$ units^2	Multiply.

Exercises

Given the coordinates of the vertices of a quadrilateral, determine whether the quadrilateral is a *square*, a *rectangle*, or a *parallelogram*. Then find the area.

1. $A(-1, 2)$, $B(3, 2)$, $C(3, -2)$, and $D(-1, -2)$

2. $R(-1, 2)$, $S(5, 0)$, $T(4, -3)$, and $U(-2, -1)$

3. $C(-2, 3)$, $D(3, 3)$, $E(5, 0)$, and $F(0, 0)$

4. $A(-2, -2)$, $B(0, 2)$, $C(4, 0)$, and $D(2, -4)$

5. $M(2, 3)$, $N(4, -1)$, $P(-2, -1)$, and $R(-4, 3)$

6. $D(2, 1)$, $E(2, -4)$, $F(-1, -4)$, and $G(-1, 1)$

11-2 Study Guide and Intervention

Areas of Triangles, Trapezoids, and Rhombi

Areas of Triangles The area of a triangle is half the area of a rectangle with the same base and height as the triangle.

If a triangle has an area of A square units, a base of b units, and a corresponding height of h units, then $A = \frac{1}{2}bh$.

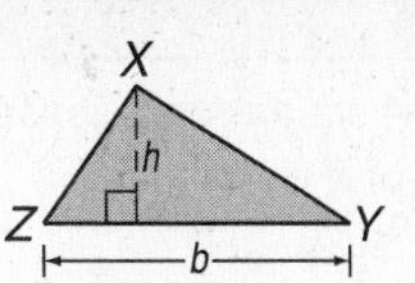

Example **Find the area of the triangle.**

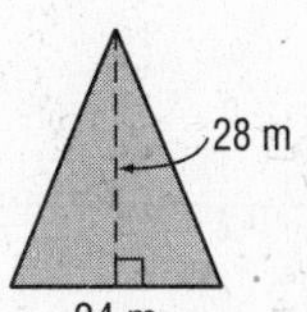

$A = \frac{1}{2}bh$	Area of a triangle
$= \frac{1}{2}(24)(28)$	$b = 24$, $h = 28$
$= 336$	Multiply.

The area is 336 square meters.

Exercises

Find the area of each figure.

1.

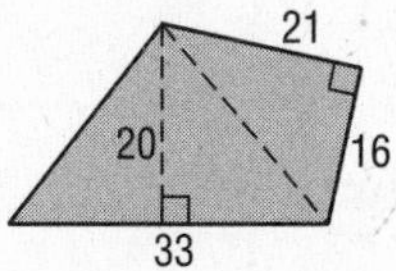

2.

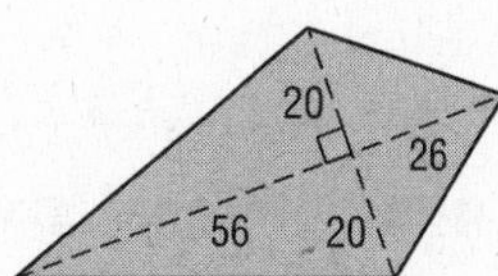

3.

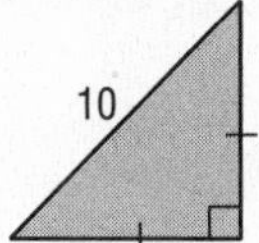

4.

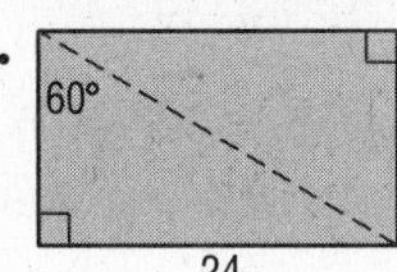

5.

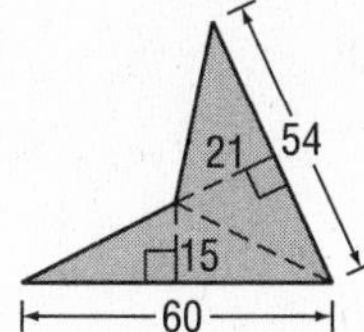

6.

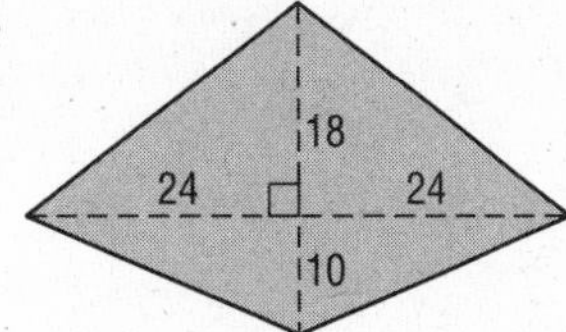

7. The area of a triangle is 72 square inches. If the height is 8 inches, find the length of the base.

8. A right triangle has a perimeter of 36 meters, a hypotenuse of 15 meters, and a leg of 9 meters. Find the area of the triangle.

11-2 Study Guide and Intervention *(continued)*

Areas of Triangles, Trapezoids, and Rhombi

Areas of Trapezoids and Rhombi The area of a trapezoid is the product of half the height and the sum of the lengths of the bases. The area of a rhombus is half the product of the diagonals.

If a trapezoid has an area of A square units, bases of b_1 and b_2 units, and a height of h units, then $A = \frac{1}{2}h(b_1 + b_2)$. 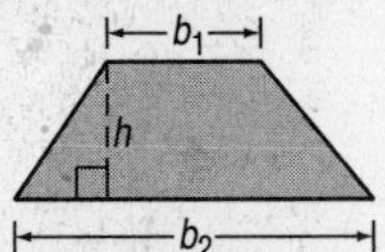	If a rhombus has an area of A square units and diagonals of d_1 and d_2 units, then $A = \frac{1}{2}d_1d_2$. 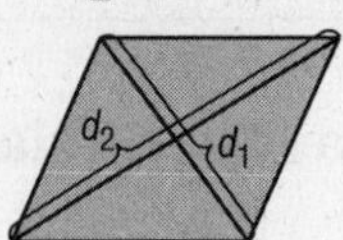

Example **Find the area of the trapezoid.**

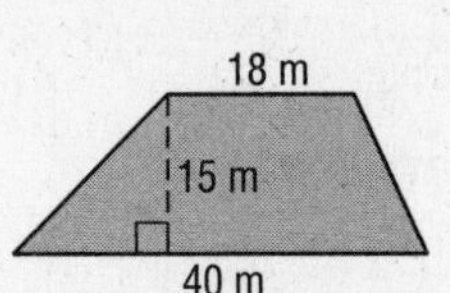

$A = \frac{1}{2}h(b_1 + b_2)$ Area of a trapezoid

$= \frac{1}{2}(15)(18 + 40)$ $h = 15$, $b_1 = 18$, $b_2 = 40$

$= 435$ Simplify.

The area is 435 square meters.

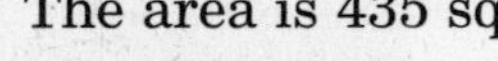

Exercises

Find the area of each quadrilateral given the coordinates of the vertices.

1.

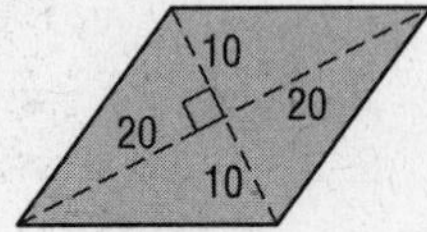

2.

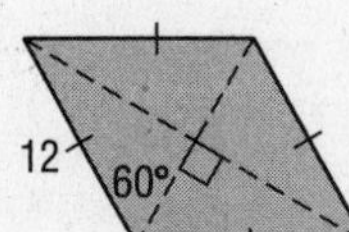

3.

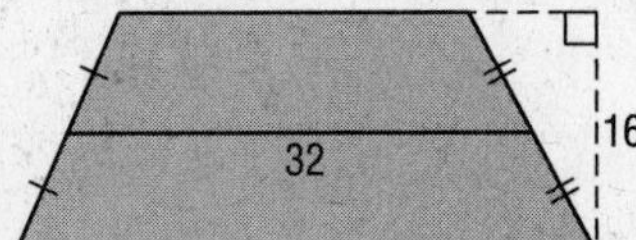

4.

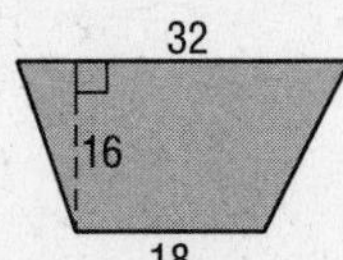

5.

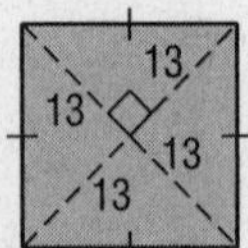

6.

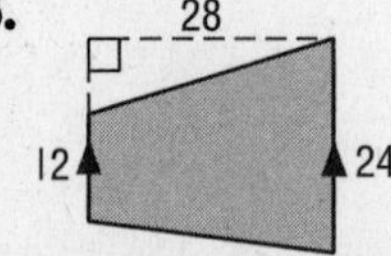

7. The area of a trapezoid is 144 square inches. If the height is 12 inches, find the length of the median.

8. A rhombus has a perimeter of 80 meters and the length of one diagonal is 24 meters. Find the area of the rhombus.

NAME ______________________________ DATE ____________ PERIOD _____

11-3 Study Guide and Intervention

Areas of Regular Polygons and Circles

Areas of Regular Polygons In a regular polygon, the segment drawn from the center of the polygon perpendicular to the opposite side is called the **apothem**. In the figure at the right, $\overline{AP}$ is the apothem and $\overline{AR}$ is the radius of the circumscribed circle.

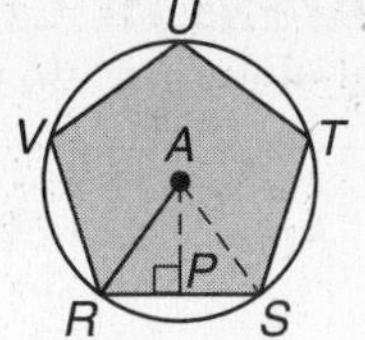

Area of a Regular Polygon	If a regular polygon has an area of A square units, a perimeter of P units, and an apothem of a units, then $A = \frac{1}{2}Pa$.

Example 1 **Verify the formula $A = \frac{1}{2}Pa$ for the regular pentagon above.**

For $\triangle RAS$, the area is $A = \frac{1}{2}bh = \frac{1}{2}(RS)(AP)$. So the area of the pentagon is $A = 5\left(\frac{1}{2}\right)(RS)(AP)$. Substituting P for $5RS$ and substituting a for AP, then $A = \frac{1}{2}Pa$.

Example 2 **Find the area of regular pentagon *RSTUV* above if its perimeter is 60 centimeters.**

First find the apothem.

The measure of central angle RAS is $\frac{360}{5}$ or 72. Therefore $m\angle RAP = 36$. The perimeter is 60, so $RS = 12$ and $RP = 6$.

$$\tan \angle RAP = \frac{RP}{AP}$$
$$\tan 36° = \frac{6}{AP}$$
$$AP = \frac{6}{\tan 36°}$$
$$\approx 8.26$$

So $A = \frac{1}{2}Pa = \frac{1}{2}60(8.26)$ or 247.7.

The area is about 248 square centimeters.

Exercises

Find the area of each regular polygon. Round to the nearest tenth.

1.

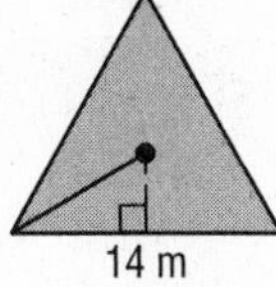

2.

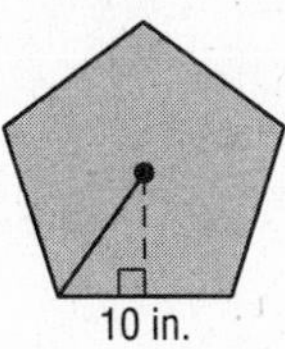

3.

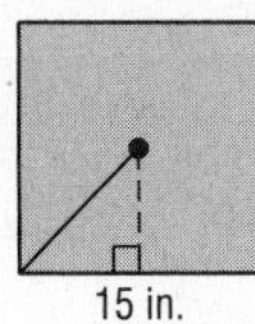

4.

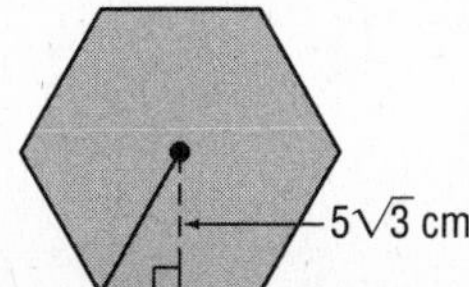

5.

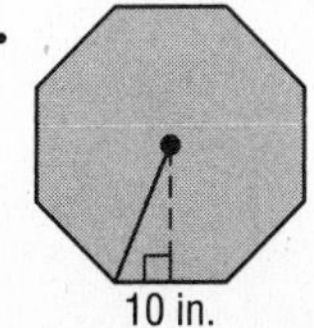

6.

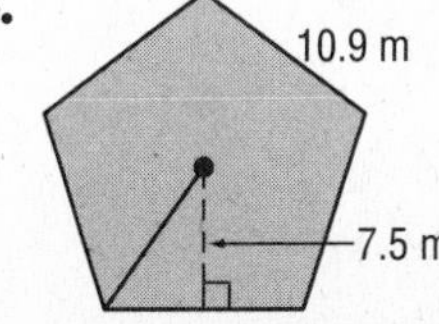

Lesson 11-3

11-3 Study Guide and Intervention *(continued)*

Areas of Regular Polygons and Circles

Areas of Circles As the number of sides of a regular polygon increases, the polygon gets closer and closer to a circle and the area of the polygon gets closer to the area of a circle.

Area of a Circle	If a circle has an area of A square units and a radius of r units, then $A = \pi r^2$. 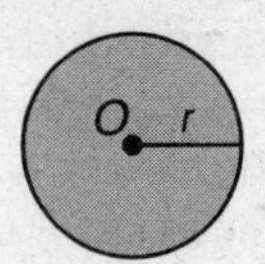

Example **Circle Q is inscribed in square $RSTU$. Find the area of the shaded region.**

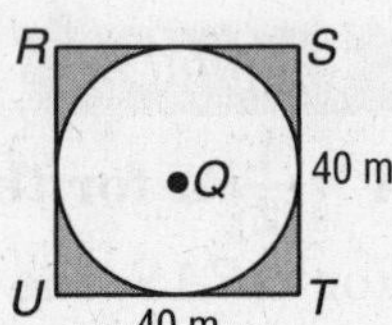

A side of the square is 40 meters, so the radius of the circle is 20 meters.

The shaded area is
Area of $RSTU$ − Area of circle Q
$= 40^2 - \pi r^2$
$= 1600 - 400\pi$
$\approx 1600 - 1256.6$
$= 343.4 \text{ m}^2$

Exercises

Find the area of each shaded region. Assume that all polygons are regular. Round to the nearest tenth.

1. 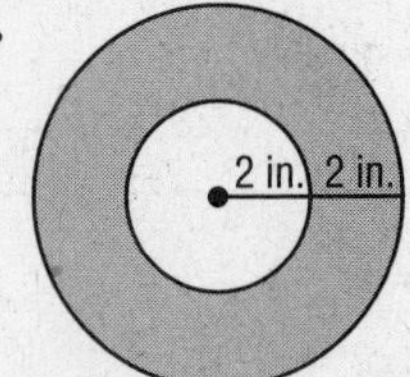

2. 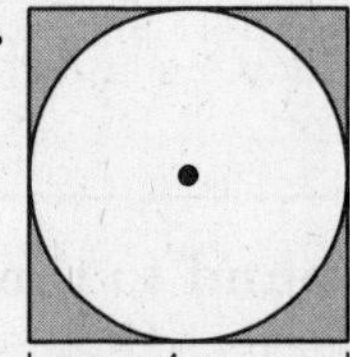

3. 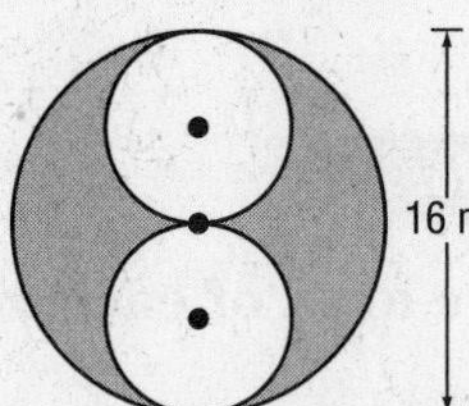

4. 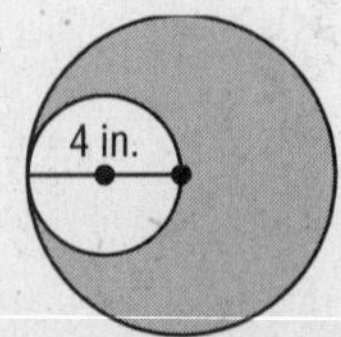

5. 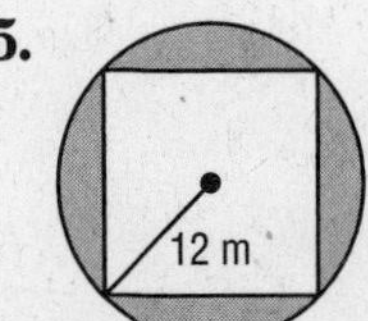

6. 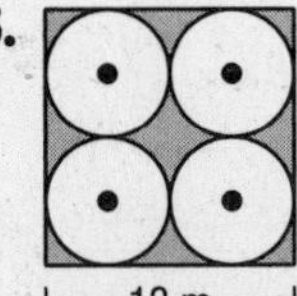

NAME ______________________ DATE ____________ PERIOD _____

11-4 Study Guide and Intervention

Areas of Irregular Figures

Irregular Figures An **irregular** figure is one that cannot be classified as one of the previously-studied shapes. To find the area of an irregular figure, break it into familiar shapes. Find the area of each shape and add the areas.

Example 1 **Find the area of the irregular figure.**

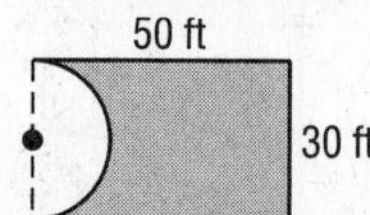

The figure is a rectangle minus one half of a circle. The radius of the circle is one half of 30 or 15.

$$A = lw - \frac{1}{2}\pi r^2$$
$$\approx 50(30) - 0.5(3.14)(15)^2$$
$$= 1146.6 \text{ or about } 1147 \text{ ft}^2$$

Example 2 **Find the area of the shaded region.**

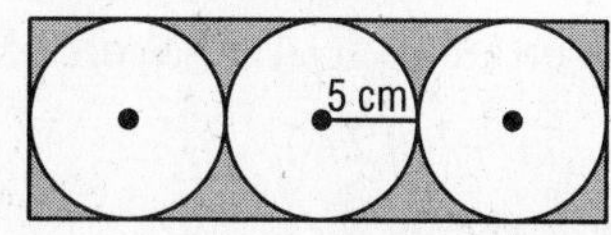

The dimensions of the rectangle are 10 centimeters and 30 centimeters. The area of the shaded region is

$$(10)(30) - 3\pi(5^2) = 300 - 75\pi$$
$$\approx 64.4 \text{ cm}^2$$

Exercises

Find the area of each figure. Round to the nearest tenth if necessary.

1. 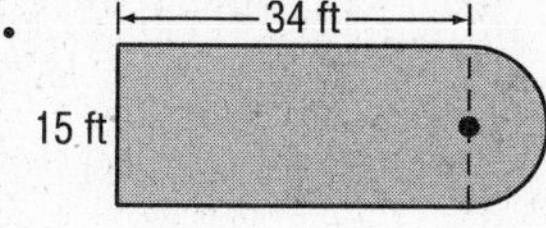

2. 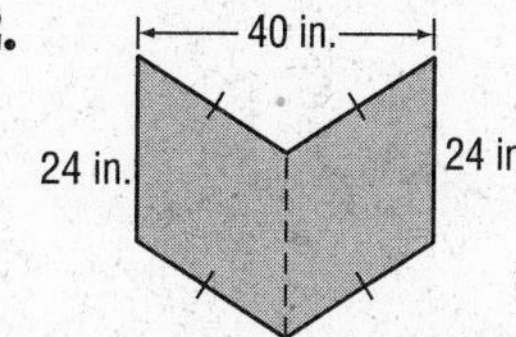

3. 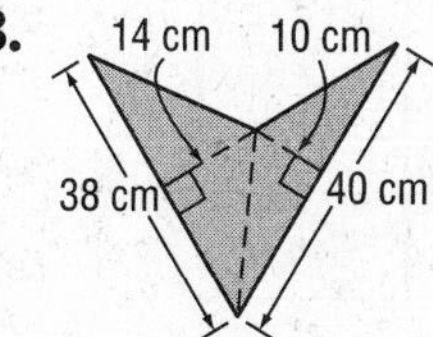

4. 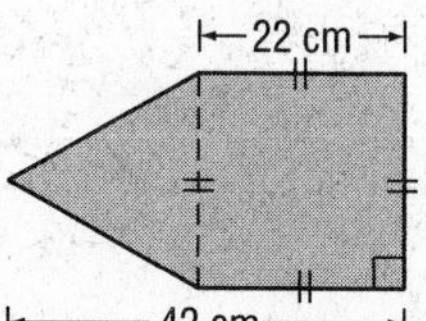

5. 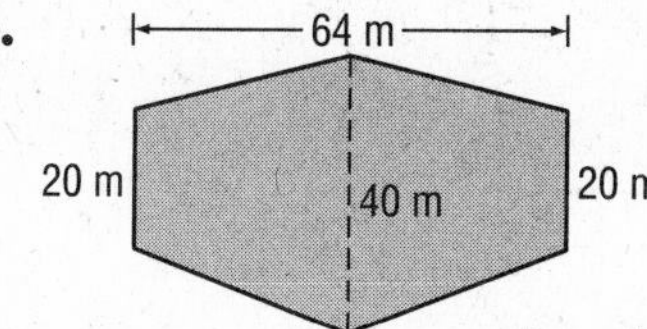

6. 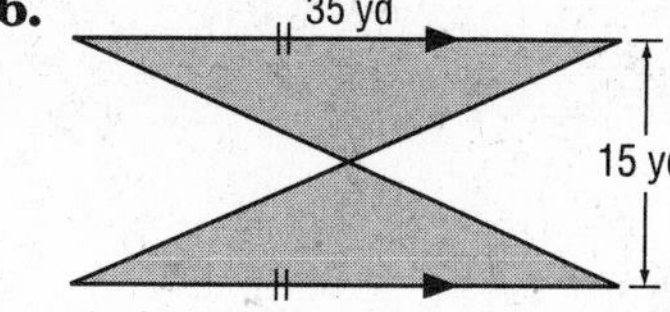

7. Refer to Example 2 above. Draw the largest possible square inside each of the three circles. What is the total area of the three squares?

Lesson 11-4

11-4 Study Guide and Intervention *(continued)*

Areas of Irregular Figures

Irregular Figures on the Coordinate Plane To find the area of an irregular figure on the coordinate plane, break up the figure into known figures. You may need to use the Distance Formula to find some of the dimensions.

Example **Find the area of irregular pentagon *ABCDE*.**

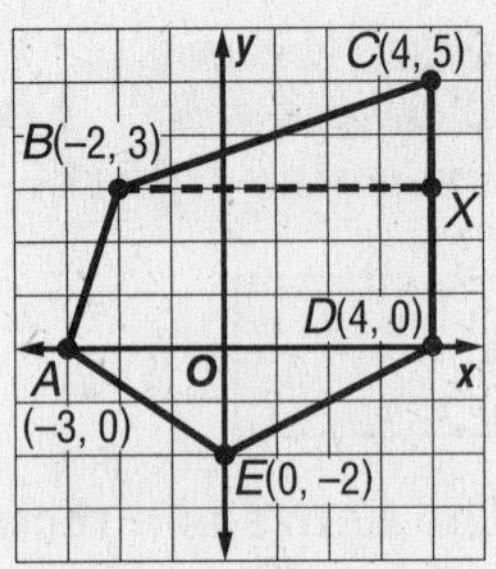

Draw $\overline{BX}$ between $B(-2, 3)$ and $X(4, 3)$ and draw $\overline{AD}$. The area of $ABCDE$ is the sum of the areas of $\triangle BCX$, trapezoid $BXDA$, and $\triangle ADE$.

A = area of $\triangle BCX$ + area of $BXDA$ + area of $\triangle ADE$

$= \frac{1}{2}(2)(6) + \frac{1}{2}(3)(6 + 7) + \frac{1}{2}(2)(7)$

$= 6 + \frac{39}{2} + 7$

$= 32.5$ square units

Exercises

Find the area of each figure. Round to the nearest tenth.

1. 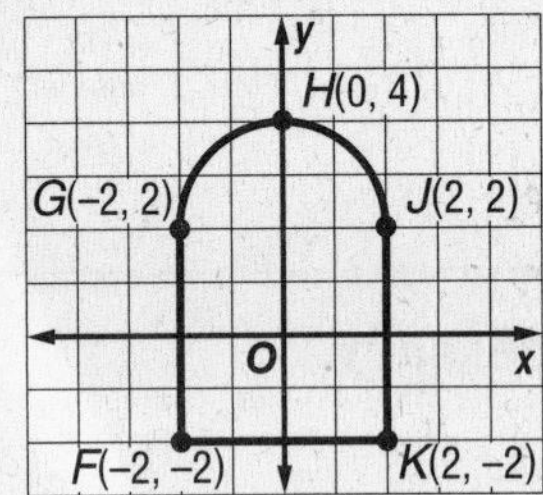

2. 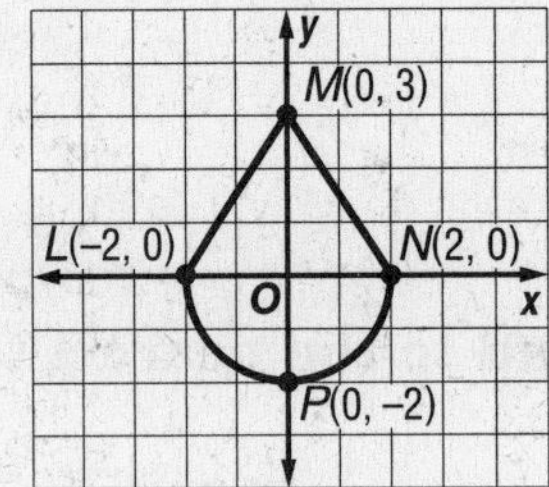

3. 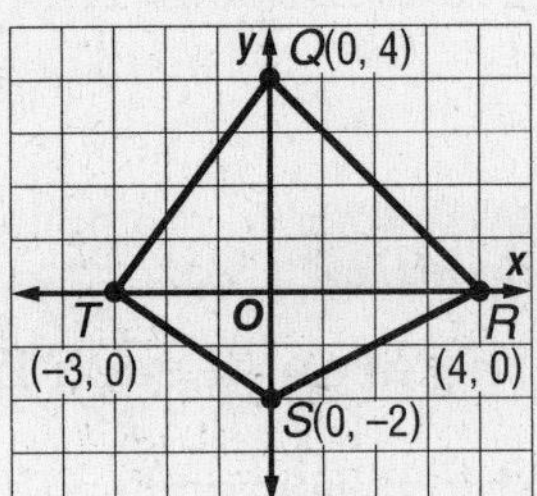

4. 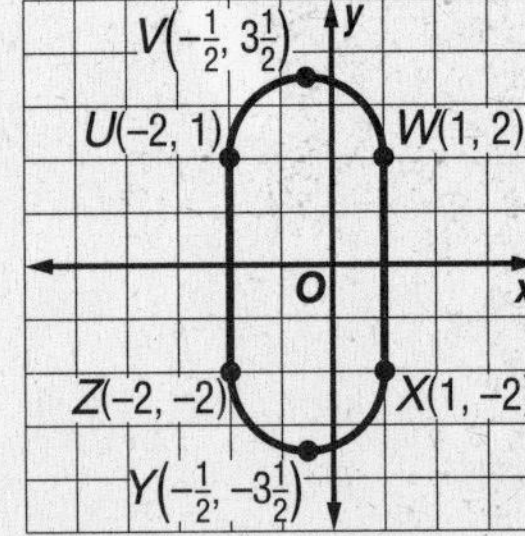

5. 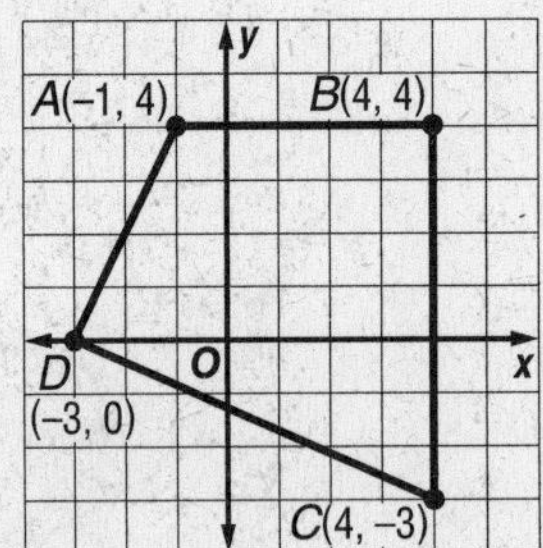

6. 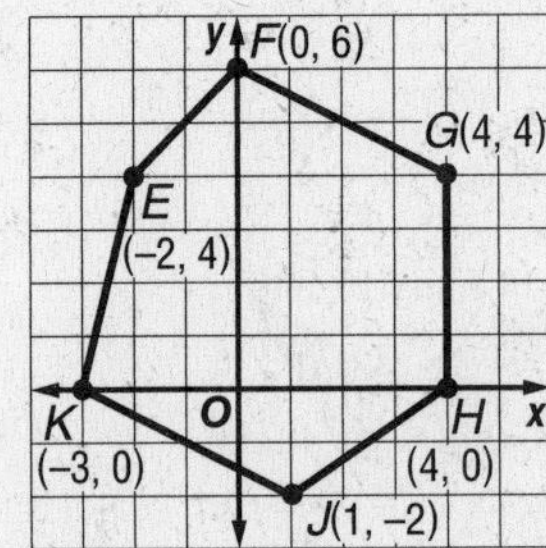

7. pentagon *ABCDE*

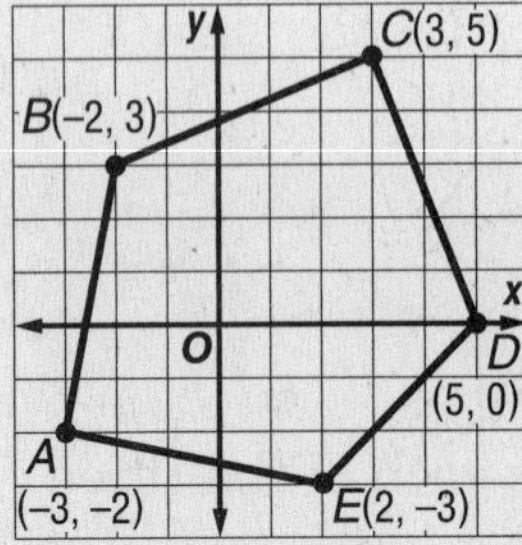

8. pentagon *RSTUV*

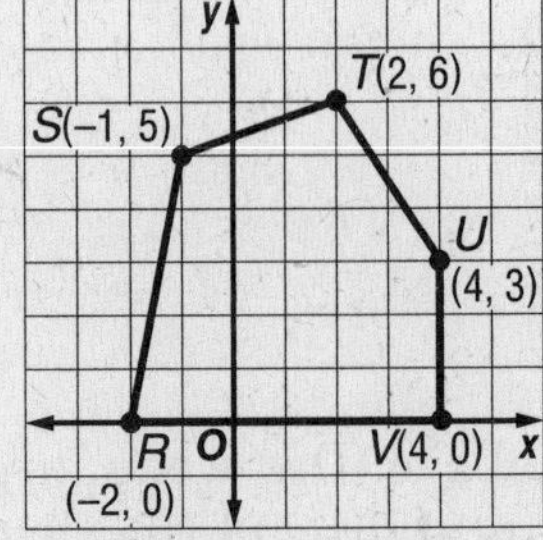

NAME ______________________________ DATE ____________ PERIOD _____

11-5 Study Guide and Intervention

Geometric Probability

Geometric Probability The probability that a point in a figure will lie in a particular part of the figure can be calculated by dividing the area of the part of the figure by the area of the entire figure. The quotient is called the **geometric probability** for the part of the figure.

> If a point in region A is chosen at random, then the probability $P(B)$ that the point is in region B, which is in the interior of region A, is
>
> $P(B) = \frac{\text{area of region } B}{\text{area of region } A}$.

Example **Darts are thrown at a circular dartboard. If a dart hits the board, what is the probability that the dart lands in the bull's-eye?**

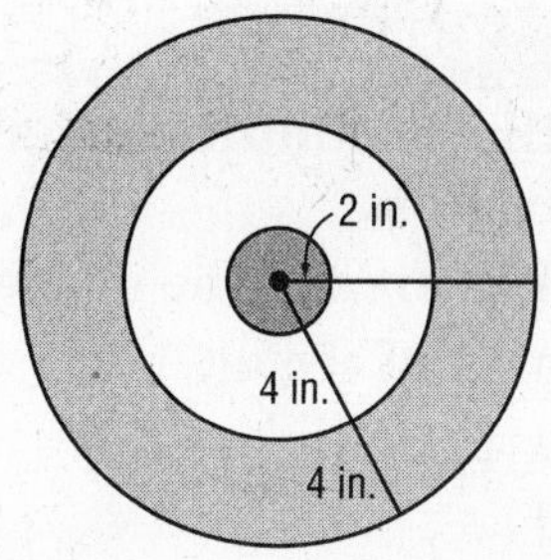

Area of bull's-eye: $A = \pi(2)^2$ or 4π

Area of entire dartboard: $A = \pi(10)^2$ or 100π

The probability of landing in the bull's-eye is

$$\frac{\text{area of bull's-eye}}{\text{area of dartboard}} = \frac{4\pi}{100\pi}$$

$$= \frac{1}{25} \text{ or } 0.04.$$

Exercises

Find the probability that a point chosen at random lies in the shaded region. Round to the nearest hundredth if necessary.

1. 

2. 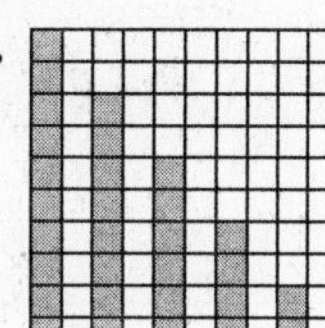

3. 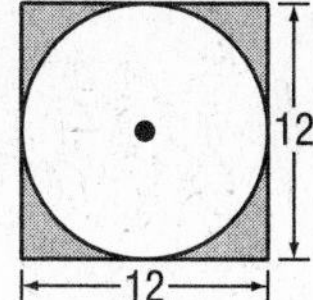

4. 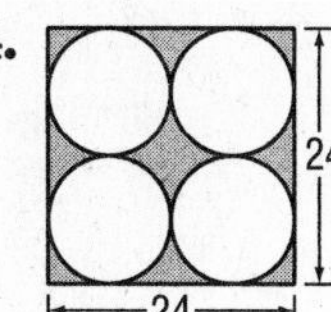

5. 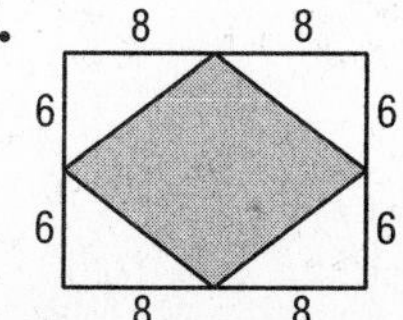

6. 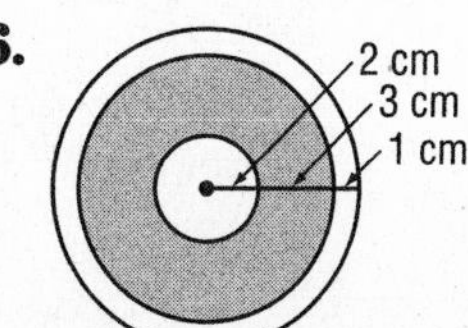

Lesson 11-5

NAME ______________________________ DATE ____________ PERIOD _____

11-5 Study Guide and Intervention *(continued)*

Geometric Probability

Sectors and Segments of Circles A **sector of a circle** is a region of a circle bounded by a central angle and its intercepted arc. A **segment of a circle** is bounded by a chord and its arc. Geometric probability problems sometimes involve sectors or segments of circles.

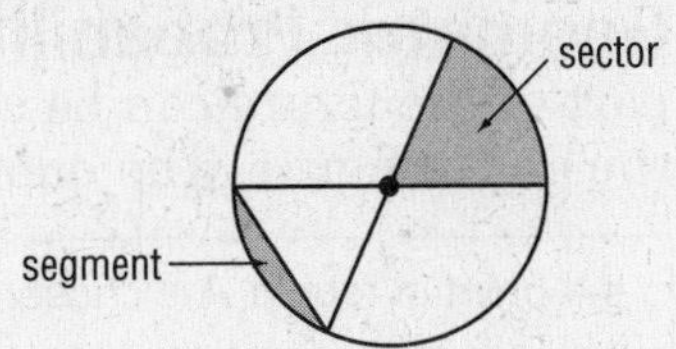

If a sector of a circle has an area of A square units, a central angle measuring $N°$, and a radius of r units, then $A = \frac{N}{360}\pi r^2$.

Example **A regular hexagon is inscribed in a circle with diameter 12. Find the probability that a point chosen at random in the circle lies in the shaded region.**

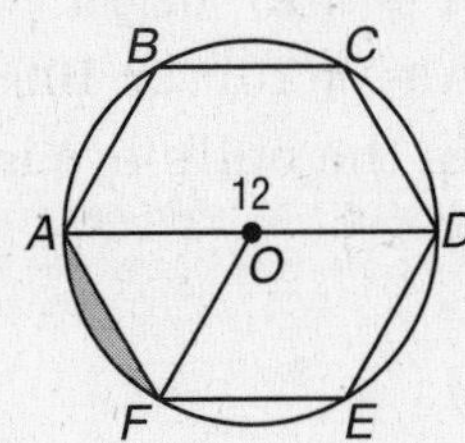

The area of the shaded segment is the area of sector AOF − the area of $\triangle AOF$.

$$\text{Area of sector } AOF = \frac{N}{360}\pi r^2 = \frac{60}{360}\pi(6^2) = 6\pi$$

$$\text{Area of } \triangle AOF = \frac{1}{2}bh = \frac{1}{2}(6)(3\sqrt{3}) = 9\sqrt{3}$$

The shaded area is $6\pi - 9\sqrt{3}$ or about 3.26.

The probability is $\frac{\text{area of segment}}{\text{area of circle}} = \frac{3.26}{36\pi}$ or about 0.03.

Exercises

Find the probability that a point in the circle chosen at random lies in the shaded region. Round to the nearest hundredth.

1. 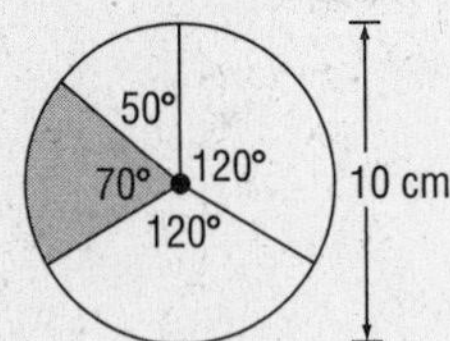

2. 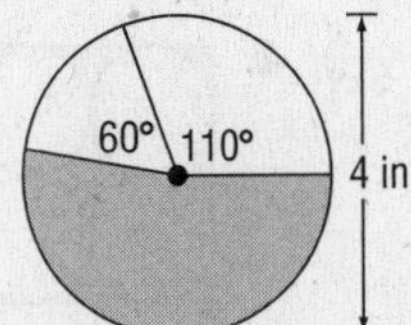

3. 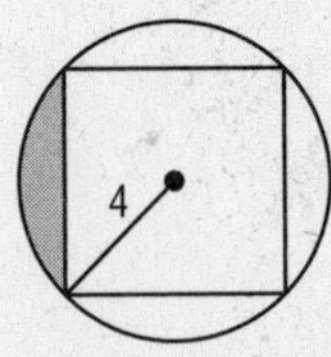

4. 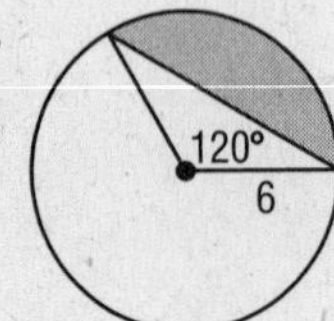

5. 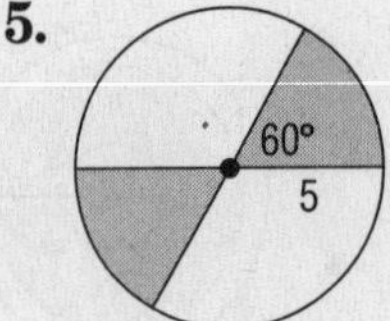

6. 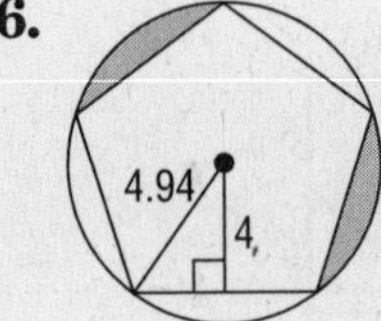

NAME ______________________ DATE __________ PERIOD _____

12-1 Study Guide and Intervention

Three-Dimensional Figures

Drawings of Three-Dimensional Figures To work with a three-dimensional object, a useful skill is the ability to make an **orthogonal drawing**, which is a set of two-dimensional drawings of the different sides of the object. For a square pyramid, you would show the top view, the left view, the front view, and the right view.

object

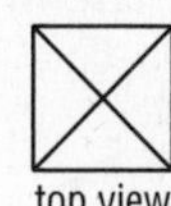
top view

left view

front view

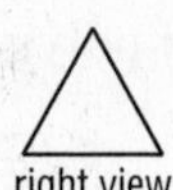
right view

Example **Draw the back view of the figure given the orthogonal drawing.**

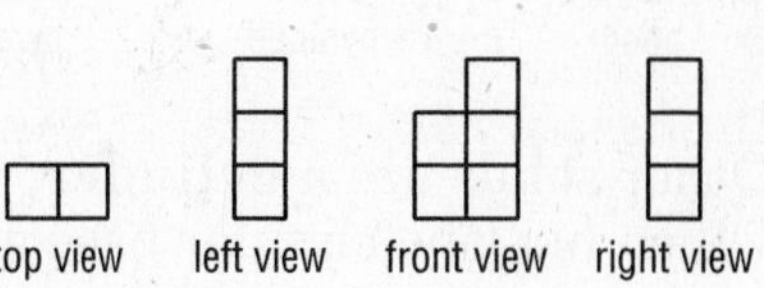
top view left view front view right view

- The top view indicates two columns.
- The left view indicates that the height of figure is three blocks.
- The front view indicates that the columns have heights 2 and 3 blocks.
- The right view indicates that the height of the figure is three blocks.

Use blocks to make a model of the object. Then use your model to draw the back view. The back view indicates that the columns have heights 3 and 2 blocks.

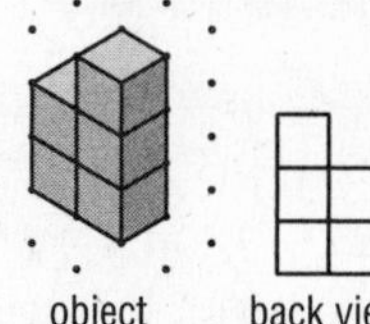
object back view

Exercises

Draw the back view and corner view of a figure given each orthogonal drawing.

1.

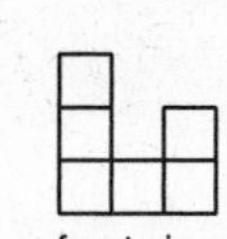

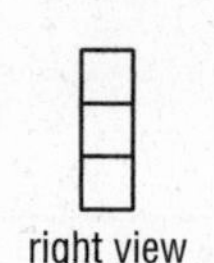

top view left view front view right view

2.

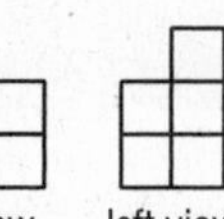

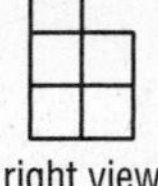

top view left view front view right view

3.

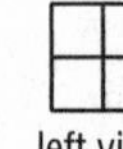

top view left view front view right view

4.

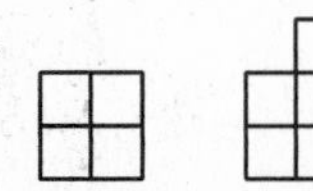

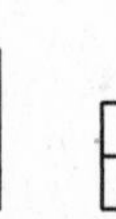

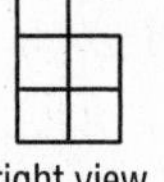

top view left view front view right view

12-1 Study Guide and Intervention *(continued)*

Three-Dimensional Figures

Identify Three-Dimensional Figures A **polyhedron** is a solid with all flat surfaces. Each surface of a polyhedron is called a **face**, and each line segment where faces intersect is called an **edge**. Two special kinds of polyhedra are **prisms**, for which two faces are congruent, parallel **bases**, and **pyramids**, for which one face is a base and all the other faces meet at a point called the **vertex**. Prisms and pyramids are named for the shape of their bases, and a regular polyhedron has a **regular** polygon as its base.

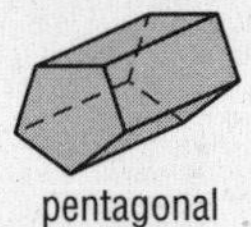
pentagonal prism

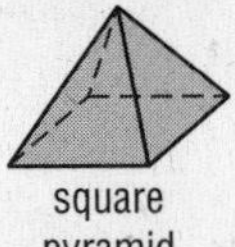
square pyramid

pentagonal pyramid

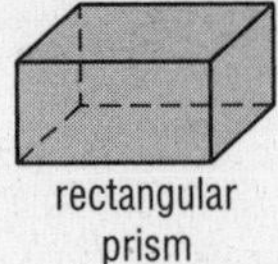
rectangular prism

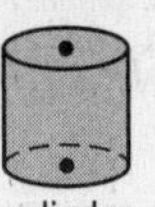
cylinder

cone

sphere

Other solids are a **cylinder**, which has congruent circular bases in parallel planes, a **cone**, which has one circular base and a vertex, and a **sphere**.

Example **Identify each solid. Name the bases, faces, edges, and vertices.**

a.

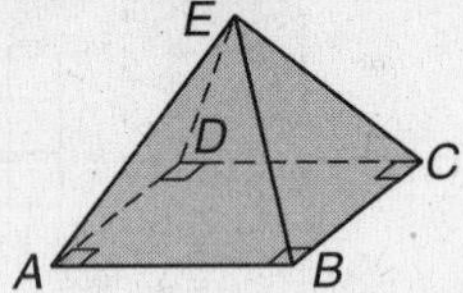

The figure is a rectangular pyramid. The base is rectangle $ABCD$, and the four faces $\triangle ABE$, $\triangle BCE$, $\triangle CDE$, and $\triangle ADE$ meet at vertex E. The edges are $\overline{AB}$, $\overline{BC}$, $\overline{CD}$, $\overline{AD}$, $\overline{AE}$, $\overline{BE}$, $\overline{CE}$, and $\overline{DE}$. The vertices are A, B, C, D, and E.

b.

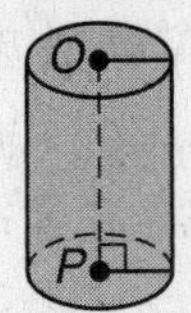

This solid is a cylinder. The two bases are $\odot O$ and $\odot P$.

Exercises

Identify each solid. Name the bases, faces, edges, and vertices.

1.

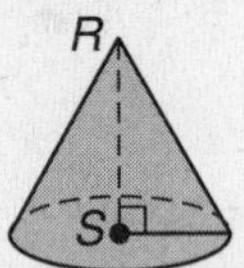

2.

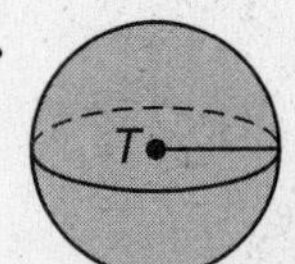

3.

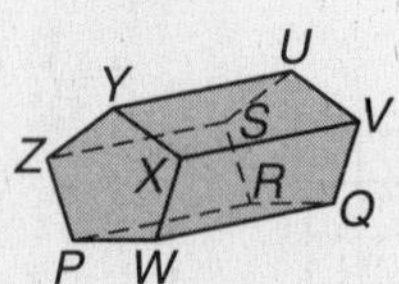

4.

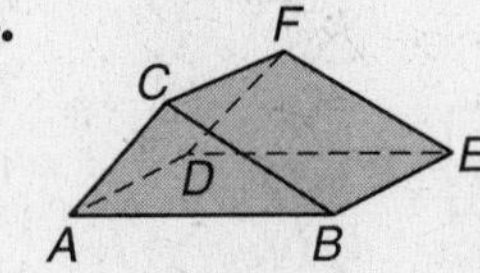

NAME ______________________ DATE __________ PERIOD ______

12-2 Study Guide and Intervention

Nets and Surface Area

Models for Three-Dimensional Figures One way to relate a three-dimensional figure and a two-dimensional drawing is to use isometric dot paper. Another way is to make a flat pattern, called a *net*, for the surfaces of a solid.

Example 1 **Use isometric dot paper to sketch a triangular prism with 3-4-5 right triangles as bases and with a height of 3 units.**

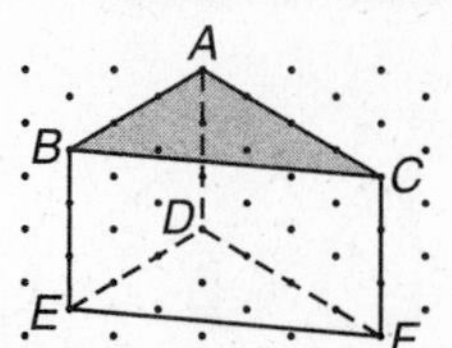

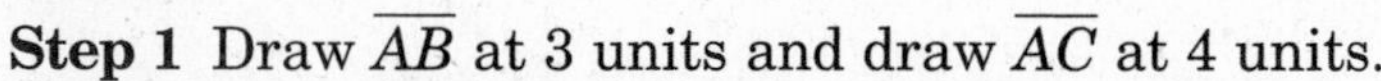

Step 1 Draw $\overline{AB}$ at 3 units and draw $\overline{AC}$ at 4 units.

Step 2 Draw $\overline{AD}$, $\overline{BE}$, and $\overline{CF}$, each at 3 units.

Step 3 Draw $\overline{BC}$ and $\triangle DEF$.

Example 2 **Match the net at the right with one of the solids below.**

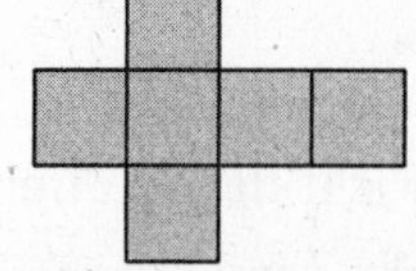

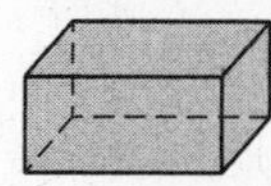
a.

b.

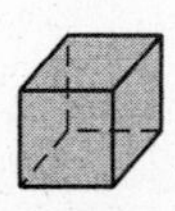
c.

The six squares of the net can be folded into a cube. The net represents solid c.

Exercises

Sketch each solid using isometric dot paper.

1. cube with edge 4

2. rectangular prism 1 unit high, 5 units long, and 4 units wide

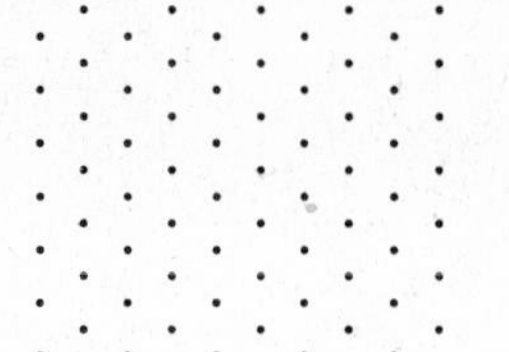

Draw a net for each solid.

3.

4.

5.

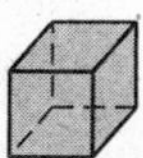

6.

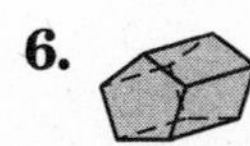

12-2 Study Guide and Intervention *(continued)*

Nets and Surface Area

Surface Area The **surface area** of a solid is the sum of the areas of the faces of the solid. Nets are useful in visualizing each face and calculating the area of the faces.

Example **Find the surface area of the triangular prism.**

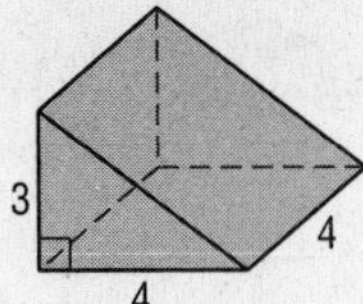

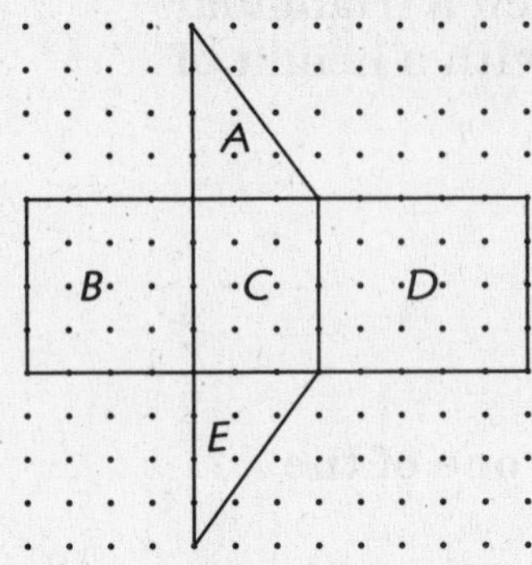

First draw a net using rectangular dot paper. Using the Pythagorean Theorem, the hypotenuse of the right triangle is $\sqrt{3^2 + 4^2}$ or 5.

$$\begin{aligned}\text{Surface area} &= A + B + C + D + E \\ &= \frac{1}{2}(4 \cdot 3) + 4 \cdot 4 + 4 \cdot 3 + 4 \cdot 5 + \frac{1}{2}(4 \cdot 3) \\ &= 60 \text{ square units}\end{aligned}$$

Exercises

Find the surface area of each solid. Round to the nearest tenth if necessary.

1.

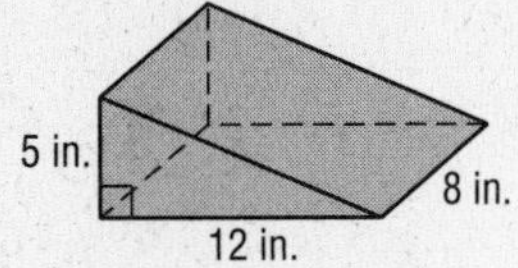

2.

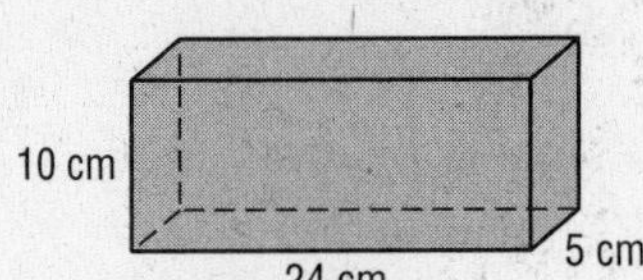

3.

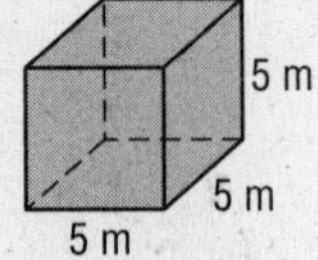

4.

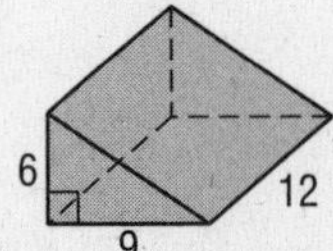

5.

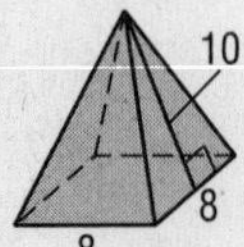

6.

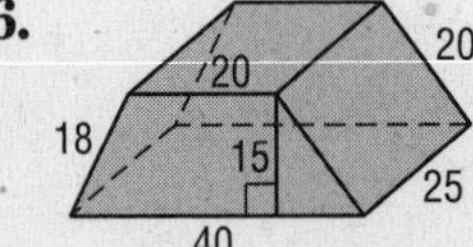

12-3 Study Guide and Intervention

Surface Areas of Prisms

Lateral Areas of Prisms Here are some characteristics of prisms.

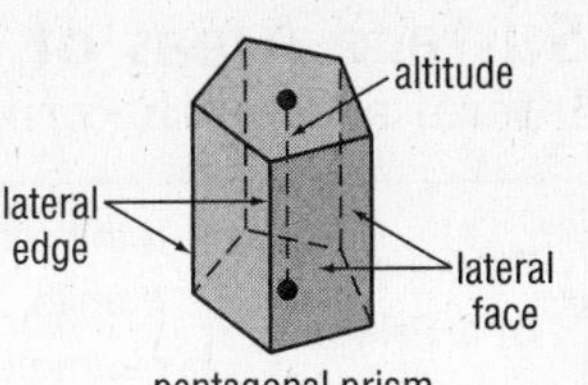

- The bases are parallel and congruent.
- The **lateral faces** are the faces that are not bases.
- The lateral faces intersect at **lateral edges**, which are parallel.
- The **altitude** of a prism is a segment that is perpendicular to the bases with an endpoint in each base.
- For a **right prism**, the lateral edges are perpendicular to the bases. Otherwise, the prism is **oblique**.

Lateral Area of a Prism	If a prism has a lateral area of L square units, a height of h units, and each base has a perimeter of P units, then $L = Ph$.

Example **Find the lateral area of the regular pentagonal prism above if each base has a perimeter of 75 centimeters and the altitude is 10 centimeters.**

$L = Ph$ Lateral area of a prism
$= 75(10)$ $P = 75$, $h = 10$
$= 750$ Multiply.

The lateral area is 750 square centimeters.

Exercises

Find the lateral area of each prism.

1.

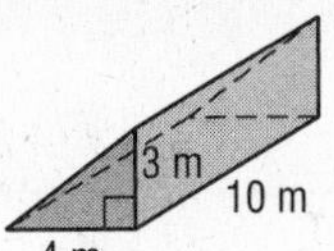

2.

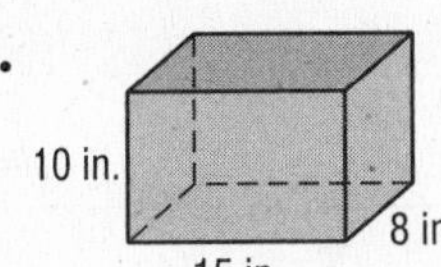

3.

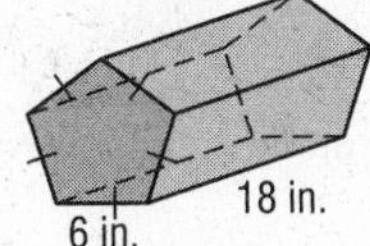

4.

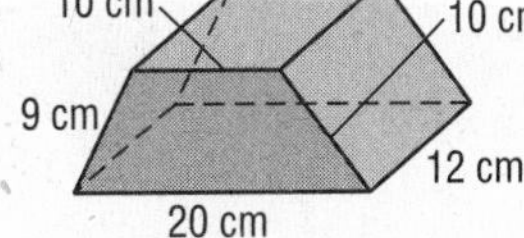

5.

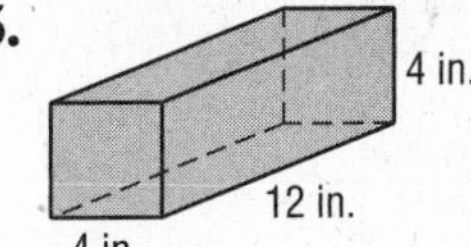

6.

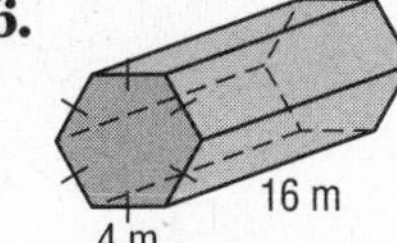

Lesson 12-3

NAME ______________________________ DATE ____________ PERIOD ______

12-3 Study Guide and Intervention *(continued)*

Surface Areas of Prisms

Surface Areas of Prisms The surface area of a prism is the lateral area of the prism plus the areas of the bases.

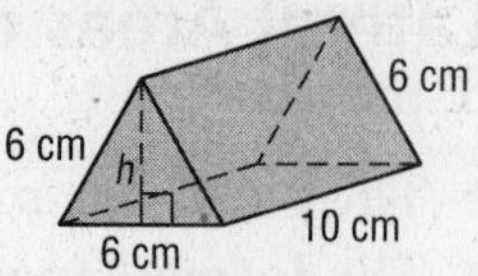

Surface Area of a Prism	If the total surface area of a prism is T square units, its height is h units, and each base has an area of B square units and a perimeter of P units, then $T = L + 2B$.

Example **Find the surface area of the triangular prism above.**

Find the lateral area of the prism.

$L = Ph$ Lateral area of a prism
$= (18)(10)$ $P = 18, h = 10$
$= 180 \text{ cm}^2$ Multiply.

Find the area of each base. Use the Pythagorean Theorem to find the height of the triangular base.

$h^2 + 3^2 = 6^2$ Pythagorean Theorem
$h^2 = 27$ Simplify.
$h = 3\sqrt{3}$ Take the square root of each side.
$B = \frac{1}{2} \times \text{base} \times \text{height}$ Area of a triangle
$= \frac{1}{2}(6)(3\sqrt{3}) \text{ or } 15.6 \text{ cm}^2$

The total area is the lateral area plus the area of the two bases.

$T = 180 + 2(15.6)$ Substitution
$= 211.2 \text{ cm}^2$ Simplify.

Exercises

Find the surface area of each prism. Round to the nearest tenth if necessary.

1.

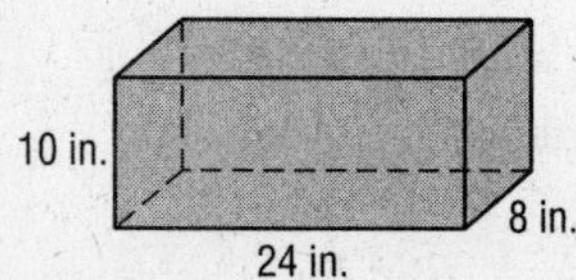

2.

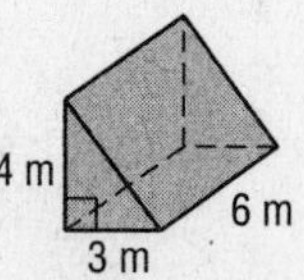

3.

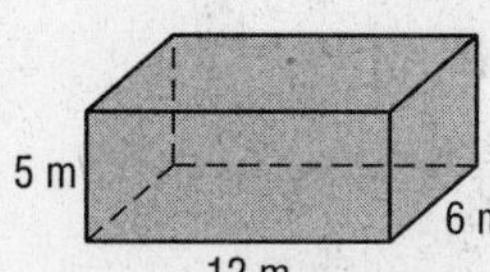

4.

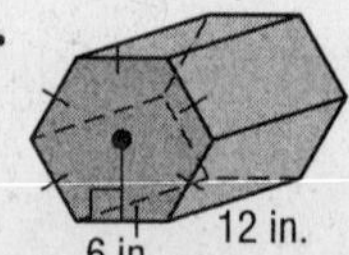

5.

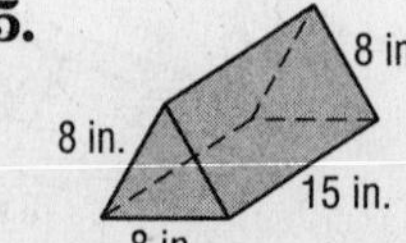

6.

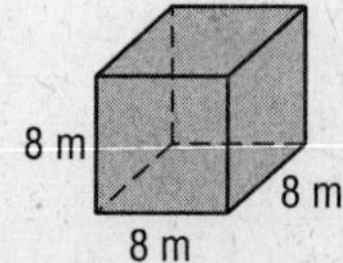

© Glencoe/McGraw-Hill

12-4 Study Guide and Intervention

Surface Areas of Cylinders

Lateral Areas of Cylinders A **cylinder** is a solid whose bases are congruent circles that lie in parallel planes. The **axis** of a cylinder is the segment whose endpoints are the centers of these circles. For a **right cylinder**, the axis and the altitude of the cylinder are equal. The lateral area of a right cylinder is the circumference of the cylinder multiplied by the height.

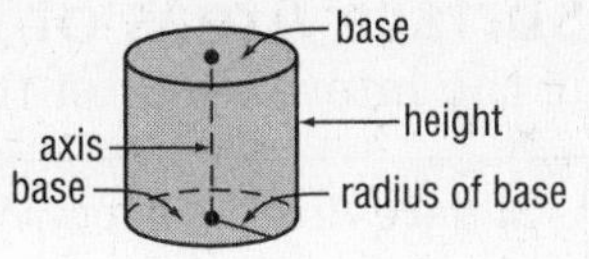

Lateral Area of a Cylinder	If a cylinder has a lateral area of L square units, a height of h units, and the bases have radii of r units, then $L = 2\pi rh$.

Example **Find the lateral area of the cylinder above if the radius of the base is 6 centimeters and the height is 14 centimeters.**

$L = 2\pi rh$ Lateral area of a cylinder

$= 2\pi(6)(14)$ Substitution

≈ 527.8 Simplify.

The lateral area is about 527.8 square centimeters.

Exercises

Find the lateral area of each cylinder. Round to the nearest tenth.

1. 4 cm, 12 cm

2. 10 in., 6 in.

3. 3 cm, 3 cm, 6 cm

4. 8 cm, 20 cm

5. 12 m, 4 m

6.

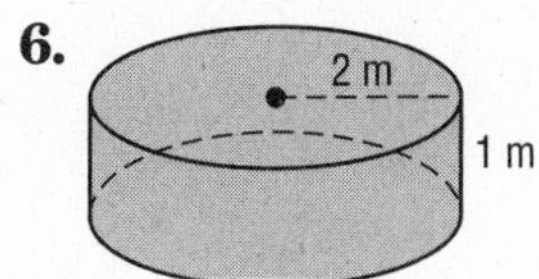

NAME ______________________________ DATE ____________ PERIOD _____

12-4 Study Guide and Intervention *(continued)*

Surface Areas of Cylinders

Surface Areas of Cylinders The surface area of a cylinder is the lateral area of the cylinder plus the areas of the bases.

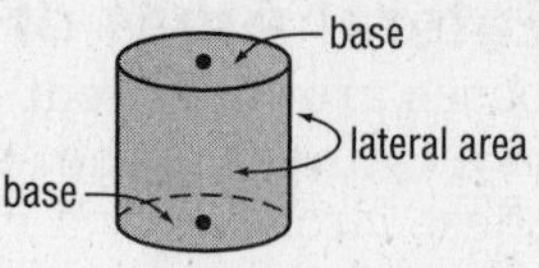

Surface Area of a Cylinder	If a cylinder has a surface area of T square units, a height of h units, and the bases have radii of r units, then $T = 2\pi rh + 2\pi r^2$.

Example **Find the surface area of the cylinder.**

Find the lateral area of the cylinder. If the diameter is 12 centimeters, then the radius is 6 centimeters.

$L = Ph$	Lateral area of a cylinder
$= (2\pi r)h$	$P = 2\pi r$
$= 2\pi(6)(14)$	$r = 6$, $h = 14$
≈ 527.8	Simplify.

Find the area of each base.

$B = \pi r^2$	Area of a circle
$= \pi(6)^2$	$r = 6$
≈ 113.1	Simplify.

The total area is the lateral area plus the area of the two bases.
$T = 527.8 + 113.1 + 113.1$ or 754 square centimeters.

Exercises

Find the surface area of each cylinder. Round to the nearest tenth.

1.

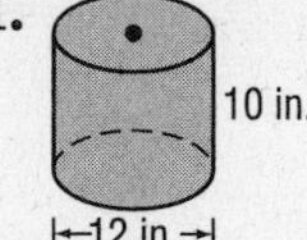

2.

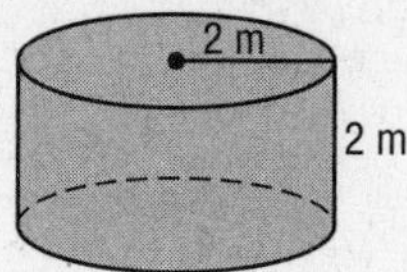

3.

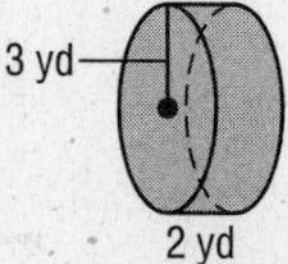

4.

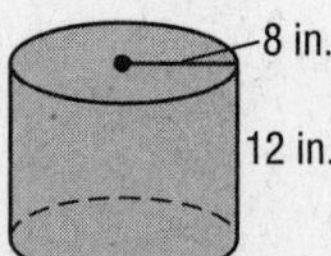

5.

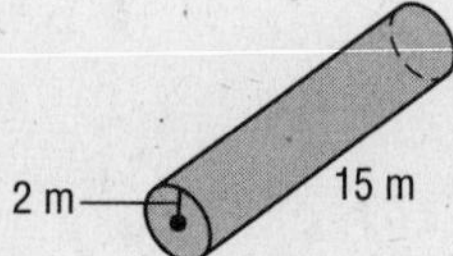

6.

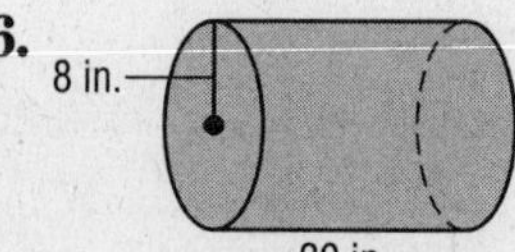

NAME ______________________ DATE ____________ PERIOD _____

12-5 Study Guide and Intervention

Surface Areas of Pyramids

Lateral Areas of Regular Pyramids Here are some properties of pyramids.

- The base is a polygon.
- All of the faces, except the base, intersect in a common point known as the **vertex**.
- The faces that intersect at the vertex, which are called **lateral faces**, are triangles.

For a **regular pyramid**, the base is a regular polygon and the **slant height** is the height of each lateral face.

Lateral Area of a Regular Pyramid	If a regular pyramid has a lateral area of L square units, a slant height of ℓ units, and its base has a perimeter of P units, then $L = \frac{1}{2}P\ell$.

Example **The roof of a barn is a regular octagonal pyramid. The base of the pyramid has sides of 12 feet, and the slant height of the roof is 15 feet. Find the lateral area of the roof.**

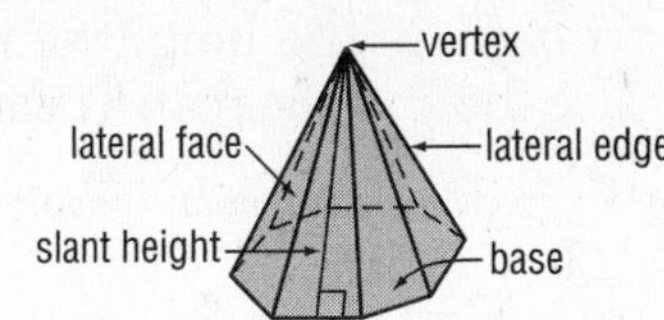

The perimeter of the base is 8(12) or 96 feet.

$L = \frac{1}{2}P\ell$ Lateral area of a pyramid

$= \frac{1}{2}(96)(15)$ $P = 96$, $\ell = 15$

$= 720$ Multiply.

The lateral area is 720 square feet.

Exercises

Find the lateral area of each regular pyramid. Round to the nearest tenth if necessary.

1.

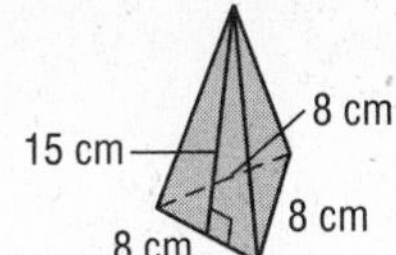

2.

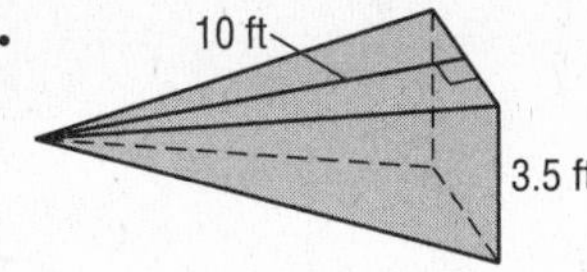

3.

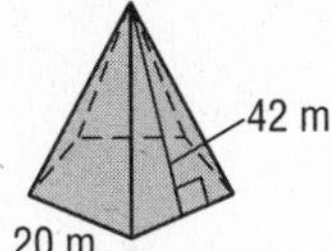

4.

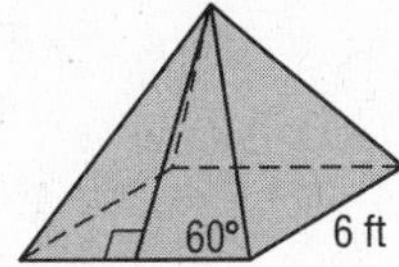

5.

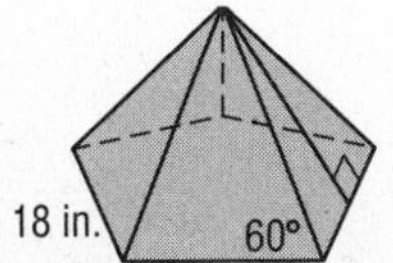

6.

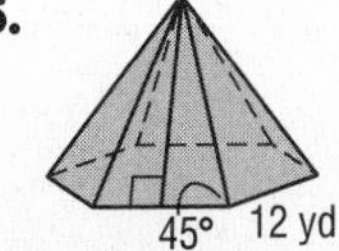

Lesson 12-5

NAME ______________________________ DATE ____________ PERIOD _____

12-5 Study Guide and Intervention *(continued)*

Surface Areas of Pyramids

Surface Areas of Regular Pyramids

The surface area of a regular pyramid is the lateral area plus the area of the base.

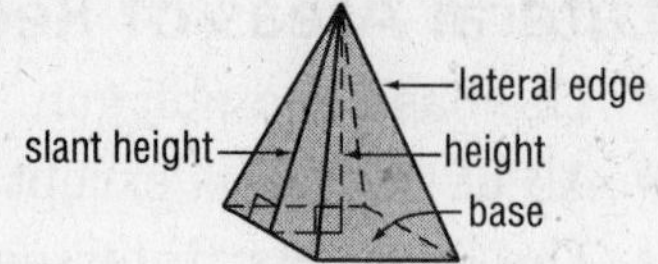

Surface Area of a Regular Pyramid	If a regular pyramid has a surface area of T square units, a slant height of ℓ units, and its base has a perimeter of P units and an area of B square units, then $T = \frac{1}{2}P\ell + B$.

Example **For the regular square pyramid above, find the surface area to the nearest tenth if each side of the base is 12 centimeters and the height of the pyramid is 8 centimeters.**

Look at the pyramid above. The slant height is the hypotenuse of a right triangle. One leg of that triangle is the height of the pyramid, and the other leg is half the length of a side of the base. Use the Pythagorean Theorem to find the slant height ℓ.

$\ell^2 = 6^2 + 8^2$ Pythagorean Theorem
$= 100$ Simplify.
$\ell = 10$ Take the square root of each side.

$T = \frac{1}{2}P\ell + B$ Surface area of a pyramid
$= \frac{1}{2}(4)(12)(10) + 12^2$ $P = (4)(12)$, $\ell = 10$, $B = 12^2$
$= 384$ Simplify.

The surface area is 384 square centimeters.

Exercises

Find the surface area of each regular pyramid. Round to the nearest tenth if necessary.

1.

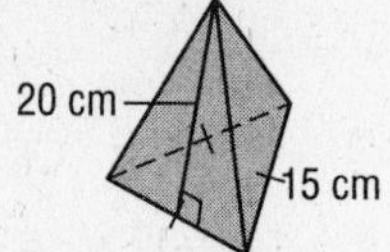

2.

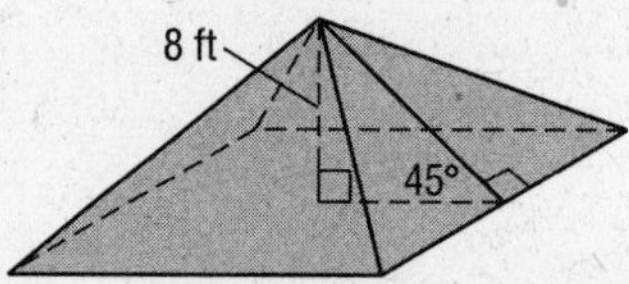

3.

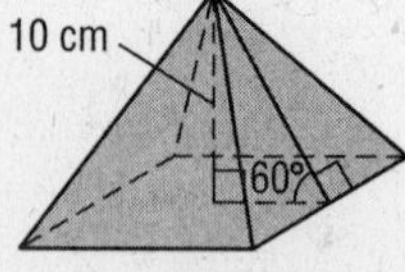

4.

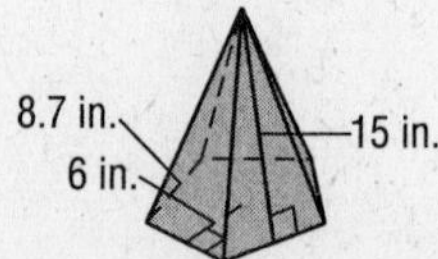

5.

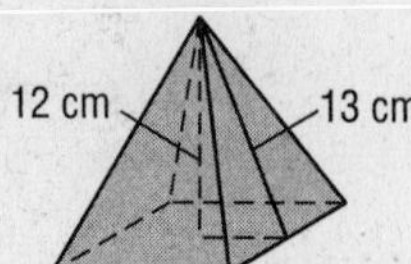

6.

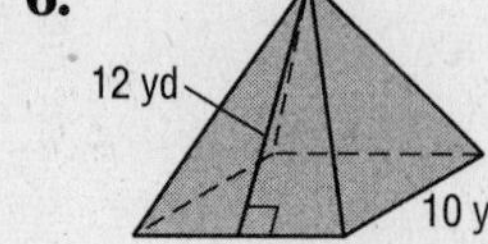

12-6 Study Guide and Intervention

Surface Areas of Cones

Lateral Areas of Cones Cones have the following properties.

- A cone has one circular base and one vertex.
- The segment whose endpoints are the vertex and the center of the base is the **axis** of the cone.
- The segment that has one endpoint at the vertex, is perpendicular to the base, and has its other endpoint on the base is the **altitude** of the cone.
- For a **right cone** the axis is also the altitude, and any segment from the circumference of the base to the vertex is the **slant height** ℓ. If a cone is not a right cone, it is oblique.

V
altitude
base
oblique cone
V
axis
ℓ
slant height
base
right cone

Lateral Area of a Cone	If a cone has a lateral area of L square units, a slant height of ℓ units, and the radius of the base is r units, then $L = \pi r\ell$.

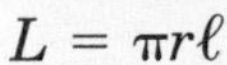

Find the lateral area of a cone with slant height of 10 centimeters and a base with a radius of 6 centimeters.

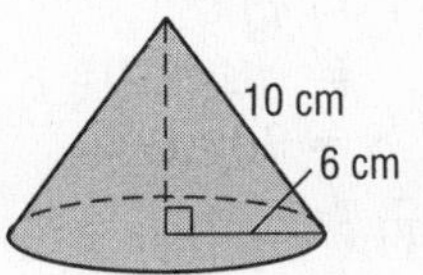

$L = \pi r\ell$ Lateral area of a cone

$= \pi(6)(10)$ $r = 6, \ell = 10$

≈ 188.5 Simplify.

The lateral area is about 188.5 square centimeters.

Exercises

Find lateral area of each circular cone. Round to the nearest tenth.

1.

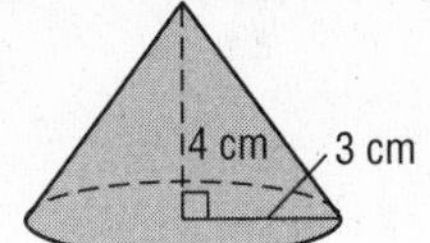

2.

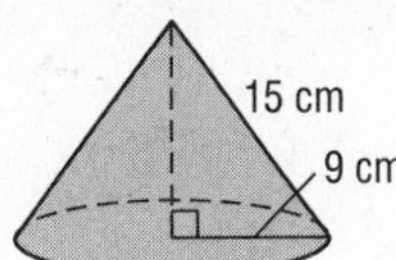

3.

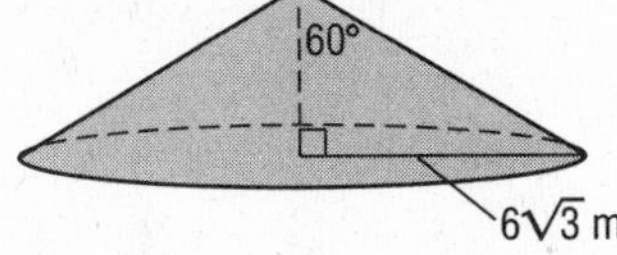

4.

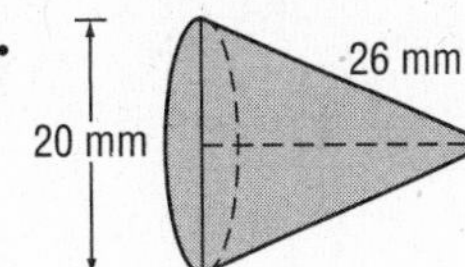

5.

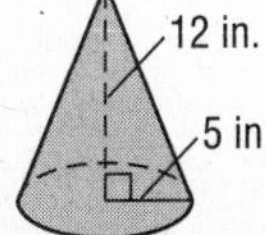

6.

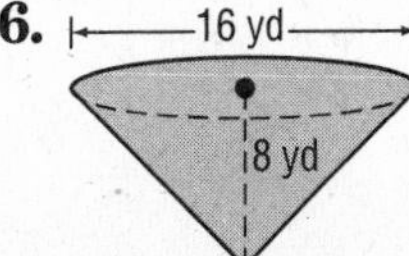

12-6 Study Guide and Intervention *(continued)*

Surface Areas of Cones

Surface Areas of Cones The surface area of a cone is the lateral area of the cone plus the area of the base.

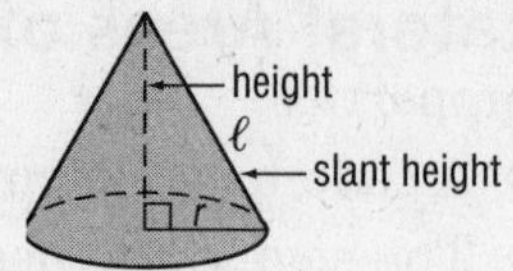

Surface Area of a Right Cone	If a cone has a surface area of T square units, a slant height of ℓ units, and the radius of the base is r units, then $T = \pi r\ell + \pi r^2$.

Example **For the cone above, find the surface area to the nearest tenth if the radius is 6 centimeters and the height is 8 centimeters.**

The slant height is the hypotenuse of a right triangle with legs of length 6 and 8. Use the Pythagorean Theorem.

$\ell^2 = 6^2 + 8^2$ Pythagorean Theorem
$\ell^2 = 100$ Simplify.
$\ell = 10$ Take the square root of each side.

$T = \pi r\ell + \pi r^2$ Surface area of a cone
$= \pi(6)(10) + \pi \cdot 6^2$ $r = 6, \ell = 10$
≈ 301.6 Simplify.

The surface area is about 301.6 square centimeters.

Exercises

Find the surface area of each cone. Round to the nearest tenth.

1.
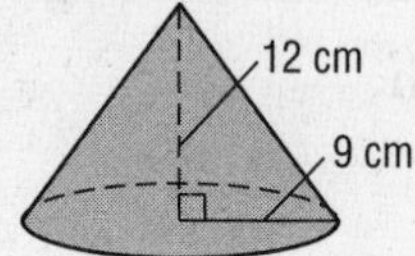

2.
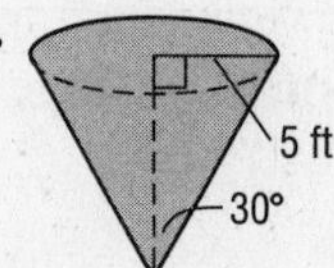

3.
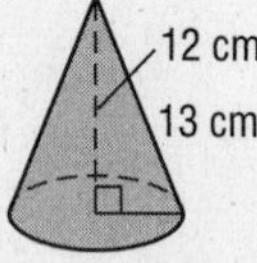

4.
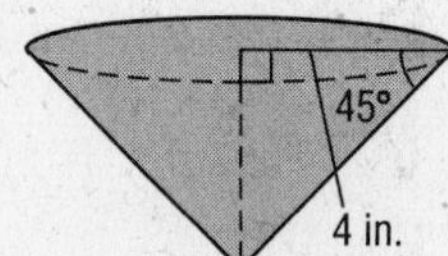

5.
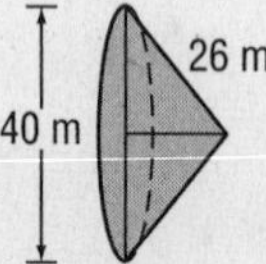

6.
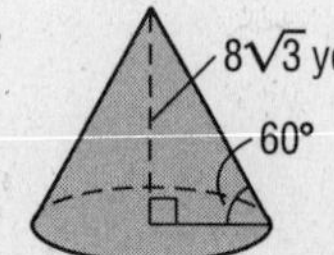

NAME ______________________ DATE ____________ PERIOD _____

12-7 Study Guide and Intervention

Surface Areas of Spheres

Properties of Spheres A **sphere** is the locus of all points that are a given distance from a given point called its **center**.

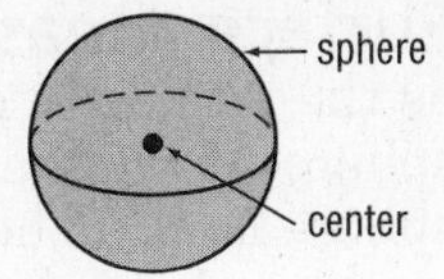

Here are some terms associated with a sphere.

- A **radius** is a segment whose endpoints are the center of the sphere and a point on the sphere.
- A **chord** is a segment whose endpoints are points on the sphere.
- A **diameter** is a chord that contains the sphere's center.
- A **tangent** is a line that intersects the sphere in exactly one point.
- A **great circle** is the intersection of a sphere and a plane that contains the center of the sphere.
- A **hemisphere** is one-half of a sphere. Each great circle of a sphere determines two hemispheres.

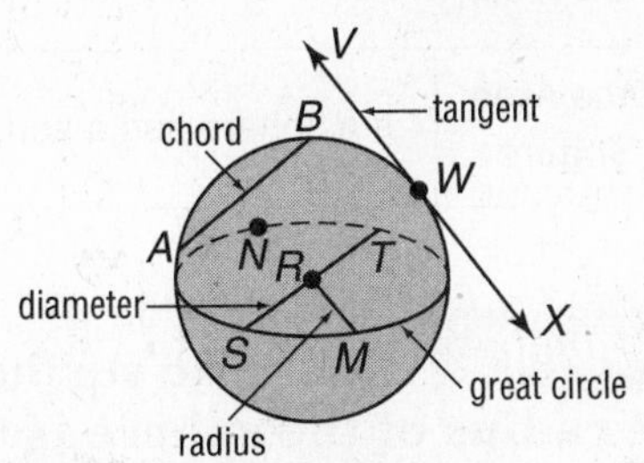

$\overline{RS}$ is a radius. $\overline{AB}$ is a chord. $\overline{ST}$ is a diameter. $\overleftrightarrow{VX}$ is a tangent. The circle that contains points S, M, T, and N is a great circle; it determines two hemispheres.

Example **Determine the shapes you get when you intersect a plane with a sphere.**

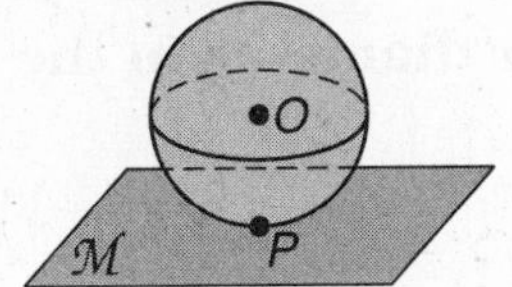

The intersection of plane $\mathcal{M}$ and sphere O is point P.

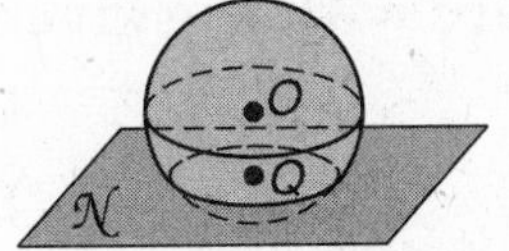

The intersection of plane $\mathcal{N}$ and sphere O is circle Q.

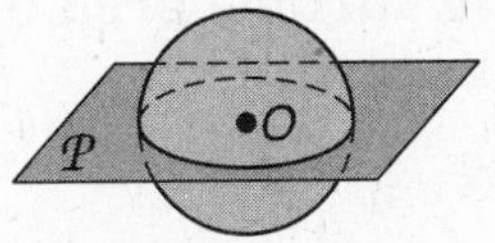

The intersection of plane $\mathcal{P}$ and sphere O is circle O.

A plane can intersect a sphere in a point, in a circle, or in a great circle.

Exercises

Describe each object as a model of a *circle*, a *sphere*, a *hemisphere*, or *none of these*.

1. a baseball

2. a pancake

3. the Earth

4. a kettle grill cover

5. a basketball rim

6. cola can

Determine whether each statement is *true* or *false*.

7. All lines intersecting a sphere are tangent to the sphere.

8. Every plane that intersects a sphere makes a great circle.

9. The eastern hemisphere of Earth is congruent to the western hemisphere.

10. The diameter of a sphere is congruent to the diameter of a great circle.

Lesson 12-7

12-7 Study Guide and Intervention *(continued)*

Surface Areas of Spheres

Surface Areas of Spheres You can think of the surface area of a sphere as the total area of all of the nonoverlapping strips it would take to cover the sphere. If r is the radius of the sphere, then the area of a great circle of the sphere is πr^2. The total surface area of the sphere is four times the area of a great circle.

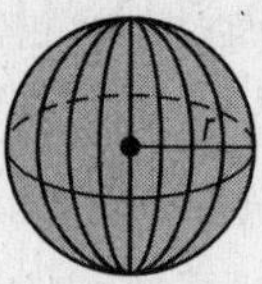

Surface Area of a Sphere	If a sphere has a surface area of T square units and a radius of r units, then $T = 4\pi r^2$.

Example **Find the surface area of a sphere to the nearest tenth if the radius of the sphere is 6 centimeters.**

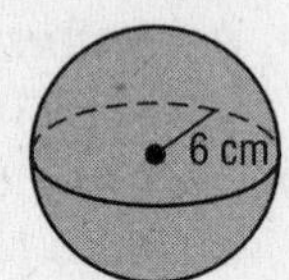

$T = 4\pi r^2$ Surface area of a sphere

$= 4\pi \cdot 6^2$ $r = 6$

≈ 452.4 Simplify.

The surface area is 452.4 square centimeters.

Exercises

Find the surface area of each sphere with the given radius or diameter to the nearest tenth.

1. $r = 8$ cm

2. $r = 2\sqrt{2}$ ft

3. $r = \pi$ cm

4. $d = 10$ in.

5. $d = 6\pi$ m

6. $d = 16$ yd

7. Find the surface area of a hemisphere with radius 12 centimeters.

8. Find the surface area of a hemisphere with diameter π centimeters.

9. Find the radius of a sphere if the surface area of a hemisphere is 192π square centimeters.

NAME ______________________ DATE ____________ PERIOD _____

13-1 Study Guide and Intervention

Volumes of Prisms and Cylinders

Volumes of Prisms The measure of the amount of space that a three-dimensional figure encloses is the **volume** of the figure. Volume is measured in units such as cubic feet, cubic yards, or cubic meters. One cubic unit is the volume of a cube that measures one unit on each edge.

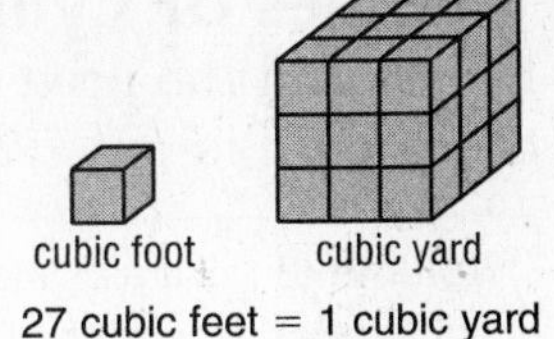

27 cubic feet = 1 cubic yard

Volume of a Prism	If a prism has a volume of V cubic units, a height of h units, and each base has an area of B square units, then $V = Bh$.

Example 1 **Find the volume of the prism.**

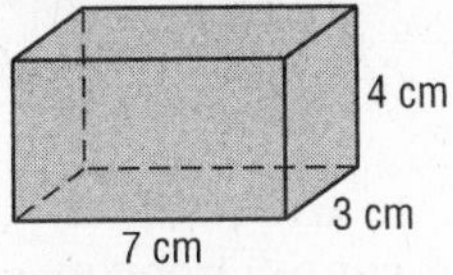

$V = Bh$ Formula for volume
$= (7)(3)(4)$ $B = (7)(3)$, $h = 4$
$= 84$ Multiply.

The volume of the prism is 84 cubic centimeters.

Example 2 **Find the volume of the prism if the area of each base is 6.3 square feet.**

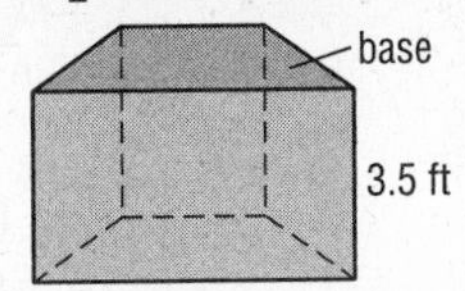

$V = Bh$ Formula for volume
$= (6.3)(3.5)$ $B = 6.3$, $h = 3.5$
$= 22.05$ Multiply.

The volume is 22.05 cubic feet.

Exercises

Find the volume of each prism. Round to the nearest tenth if necessary.

1.

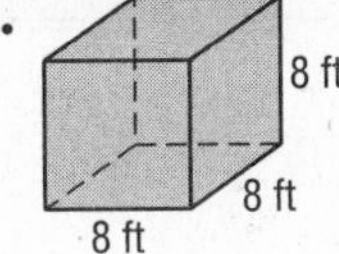

2.

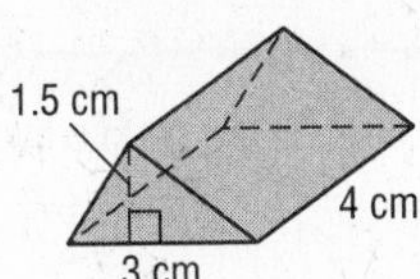

3.

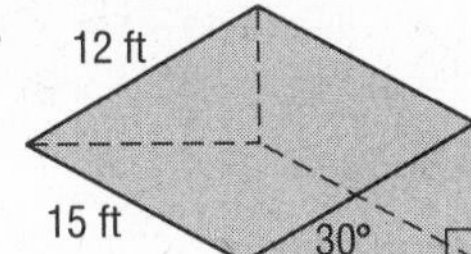

4.

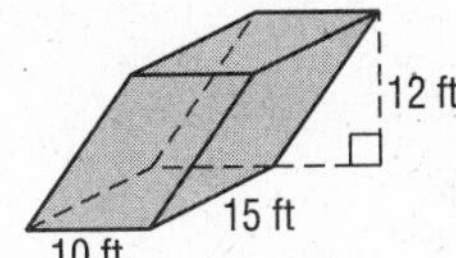

5.

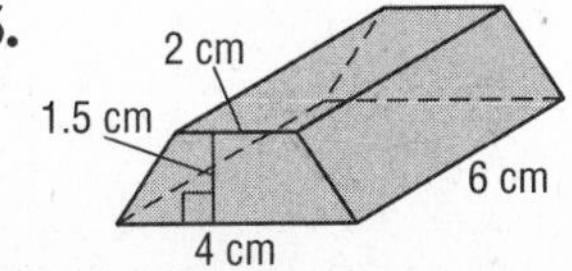

6.

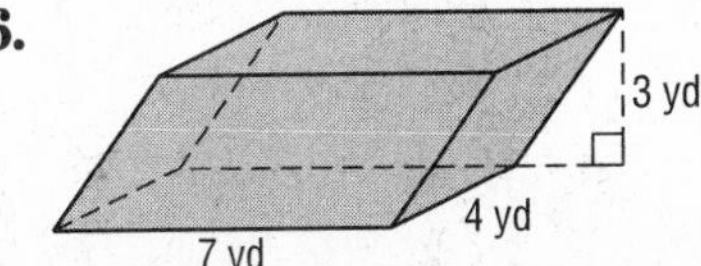

NAME _______________ DATE _______ PERIOD _____

13-1 Study Guide and Intervention *(continued)*

Volumes of Prisms and Cylinders

Volumes of Cylinders The volume of a cylinder is the product of the height and the area of the base. The base of a cylinder is a circle, so the area of the base is πr^2.

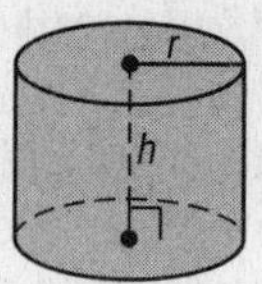

Volume of a Cylinder	If a cylinder has a volume of V cubic units, a height of h units, and the bases have radii of r units, then $V = \pi r^2 h$.

Example 1 **Find the volume of the cylinder.**

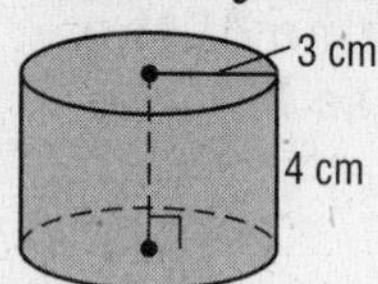

$V = \pi r^2 h$ Volume of a cylinder

$= \pi(3)^2(4)$ $r = 3$, $h = 4$

≈ 113.1 Simplify.

The volume is about 113.1 cubic centimeters.

Example 2 **Find the area of the oblique cylinder.**

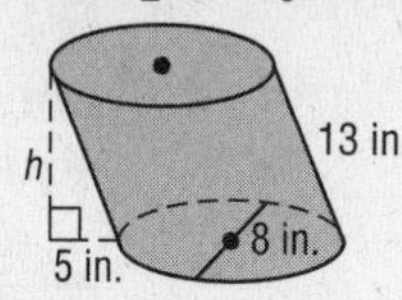

The radius of each base is 4 inches, so the area of the base is 16π in^2. Use the Pythagorean Theorem to find the height of the cylinder.

$h^2 + 5^2 = 13^2$ Pythagorean Theorem

$h^2 = 144$ Simplify.

$h = 12$ Take the square root of each side.

$V = \pi r^2 h$ Volume of a cylinder

$= \pi(4)^2(12)$ $r = 4$, $h = 12$

≈ 603.2 in^3 Simplify.

Exercises

Find the volume of each cylinder. Round to the nearest tenth.

1.

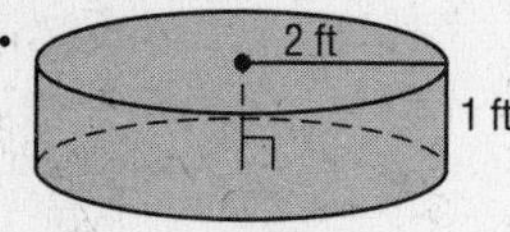

2.

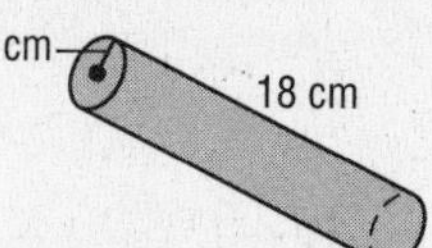

3.

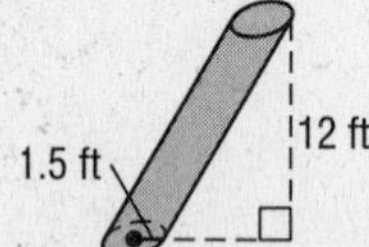

4.
20 ft
20 ft

5.

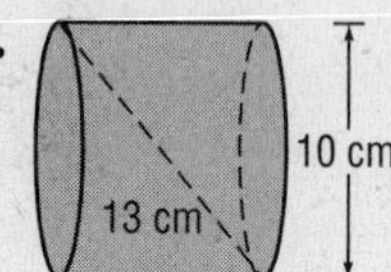

6.

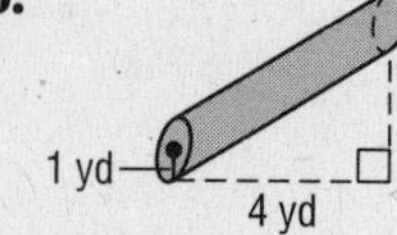

NAME ____________________ DATE ____________ PERIOD _____

13-2 Study Guide and Intervention

Volumes of Pyramids and Cones

Volumes of Pyramids This figure shows a prism and a pyramid that have the same base and the same height. It is clear that the volume of the pyramid is less than the volume of the prism. More specifically, the volume of the pyramid is one-third of the volume of the prism.

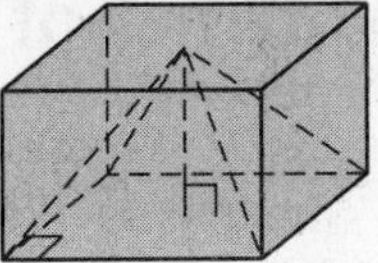

Volume of a Pyramid	If a pyramid has a volume of V cubic units, a height of h units, and a base with an area of B square units, then $V = \frac{1}{3}Bh$.

Example **Find the volume of the square pyramid.**

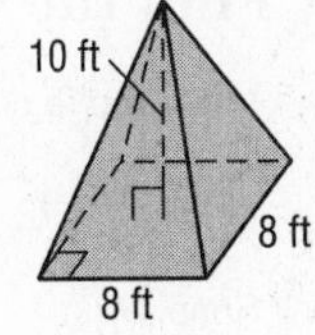

$V = \frac{1}{3}Bh$ Volume of a pyramid

$= \frac{1}{3}(8)(8)10$ $B = (8)(8)$, $h = 10$

≈ 213.3 Multiply.

The volume is about 213.3 cubic feet.

Exercises

Find the volume of each pyramid. Round to the nearest tenth if necessary.

1.

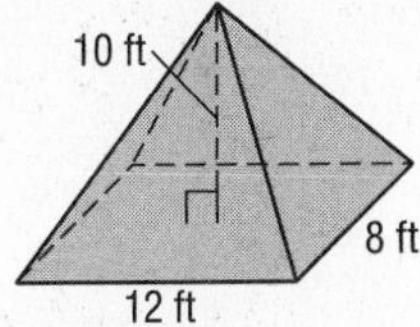

2.

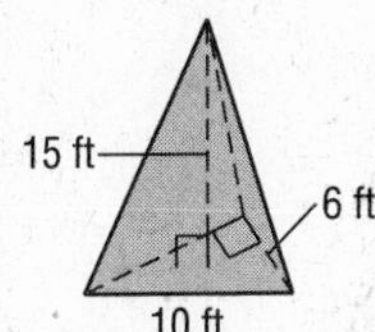

3.

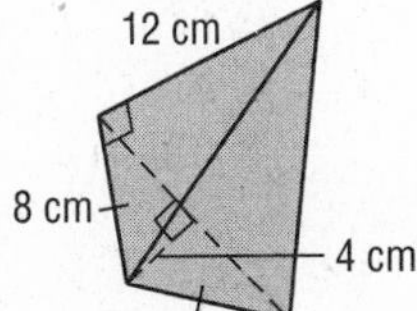

4.

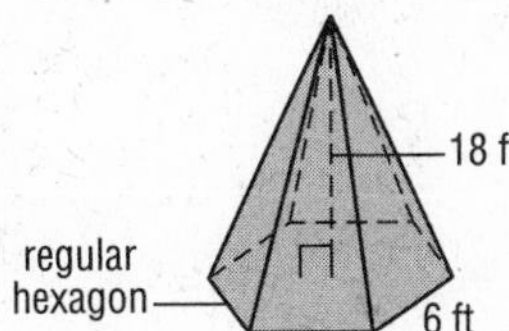

5.

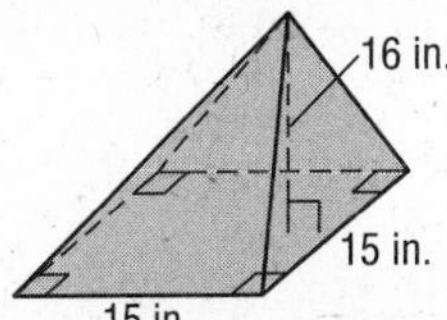

6.

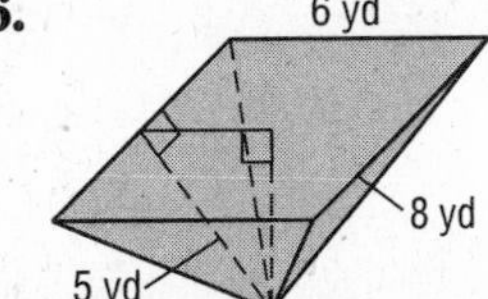

Lesson 13-2

13-2 Study Guide and Intervention *(continued)*

Volumes of Pyramids and Cones

Volumes of Cones For a cone, the volume is one-third the product of the height and the base. The base of a cone is a circle, so the area of the base is πr^2.

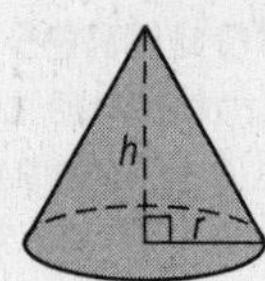

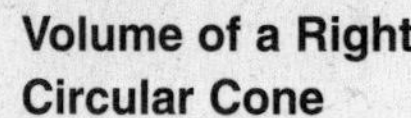

Volume of a Right Circular Cone	If a cone has a volume of V cubic units, a height of h units, and the area of the base is B square units, then $V = \frac{1}{3}Bh$.

The same formula can be used to find the volume of oblique cones.

Example **Find the volume of the cone.**

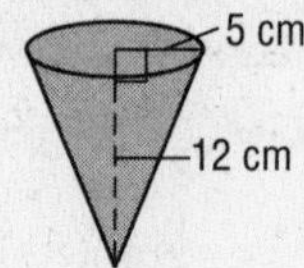

$V = \frac{1}{3}\pi r^2 h$ Volume of a cone

$= \frac{1}{3}\pi(5)^2 12$ $r = 5$, $h = 12$

≈ 314.2 Simplify.

The volume of the cone is about 314.2 cubic centimeters.

Exercises

Find the volume of each cone. Round to the nearest tenth.

1.

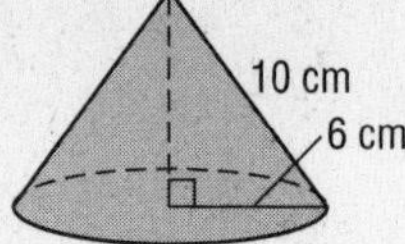

2.

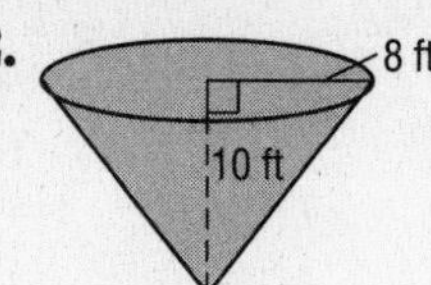

3.

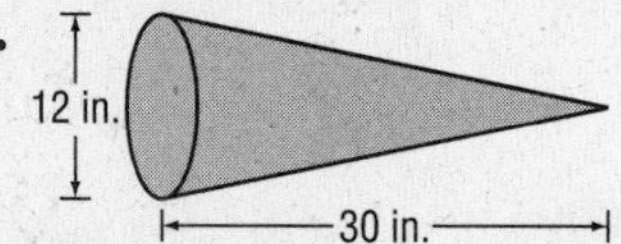

4.

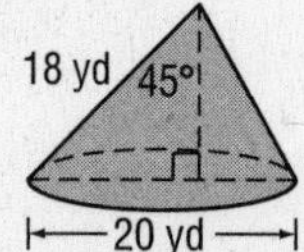

5.

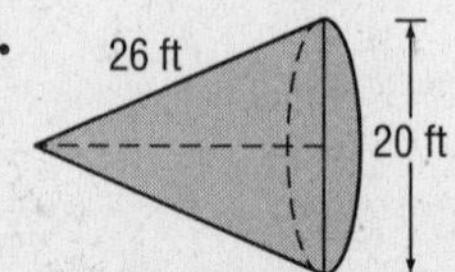

6.

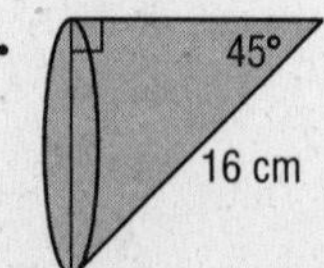

13-3 Study Guide and Intervention

Volumes of Spheres

Volumes of Spheres A sphere has one basic measurement, the length of its radius. If you know the radius of a sphere, you can calculate its volume.

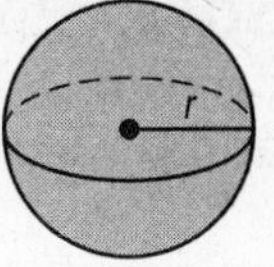

Volume of a Sphere	If a sphere has a volume of V cubic units and a radius of r units, then $V = \frac{4}{3}\pi r^3$.

Example 1 **Find the volume of a sphere with radius 8 centimeters.**

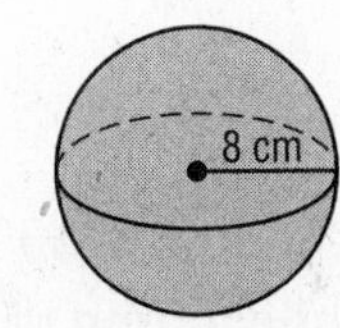

$V = \frac{4}{3}\pi r^3$ Volume of a sphere

$= \frac{4}{3}\pi(8)^3$ $r = 8$

≈ 2144.7 Simplify.

The volume is about 2144.7 cubic centimeters.

Example 2 **A sphere with radius 5 inches just fits inside a cylinder. What is the difference between the volume of the cylinder and the volume of the sphere? Round to the nearest cubic inch.**

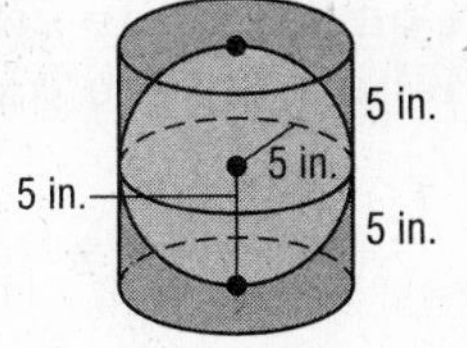

The base of the cylinder is 25π in^2 and the height is 10 in., so the volume of the cylinder is 250π in^3. The volume of the sphere is $\frac{4}{3}\pi(5)^3$ or $\frac{500\pi}{3}$ in^3. The difference in the volumes is $250\pi - \frac{500\pi}{3}$ or about 262 in^3.

Exercises

Find the volume of each solid. Round to the nearest tenth.

1.
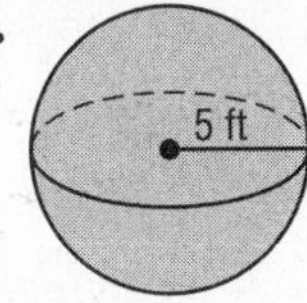

2.
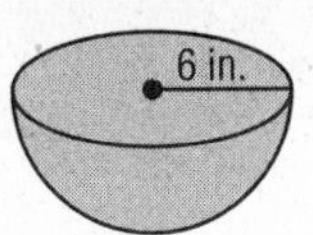

3.
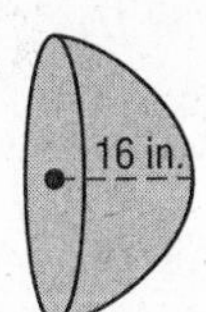

4.
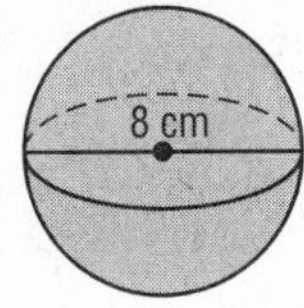

5.
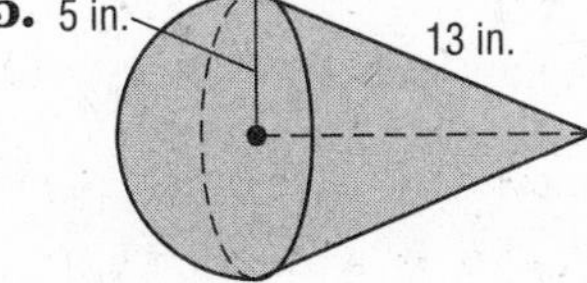

6.
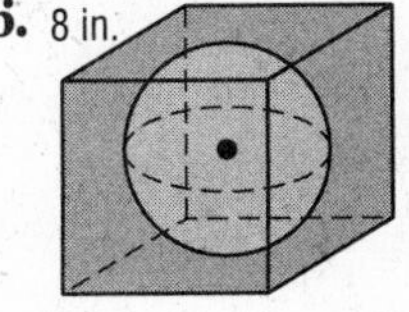

7. A hemisphere with radius 16 centimeters just fits inside a rectangular prism. What is the difference between the volume of the prism and the volume of the hemisphere? Round to the nearest cubic centimeter.

Lesson 13-3

13-3 Study Guide and Intervention *(continued)*

Volumes of Spheres

Solve Problems Involving Volumes of Spheres If you want to know if a sphere can be packed inside another container, or if you want to compare the capacity of a sphere and another shape, you can compare volumes.

Example **Compare the volumes of the sphere and the cylinder. Determine which quantity is greater.**

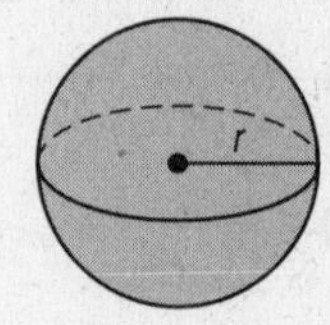

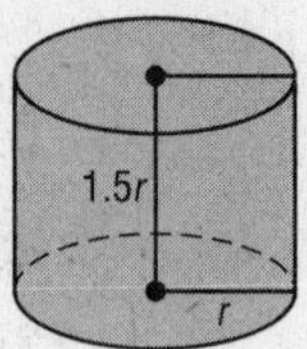

$V = \frac{4}{3}\pi r^3$ Volume of sphere

$V = \pi r^2 h$ Volume of cylinder
$= \pi r^2(1.5r)$ $h = 1.5r$
$= 1.5\pi r^3$ Simplify.

Compare $\frac{4}{3}\pi r^3$ with $1.5\pi r^3$. Since $\frac{4}{3}$ is less than 1.5, it follows that the volume of the sphere is less than the volume of the cylinder.

Exercises

Compare the volume of a sphere with radius r to the volume of each figure below. Which figure has a greater volume?

1.

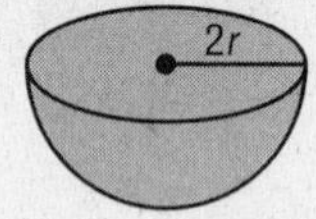

2.

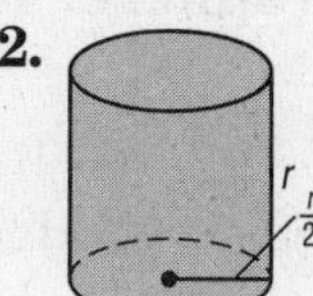

3.

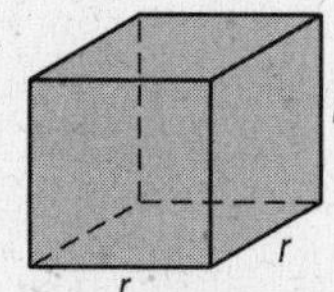

4.

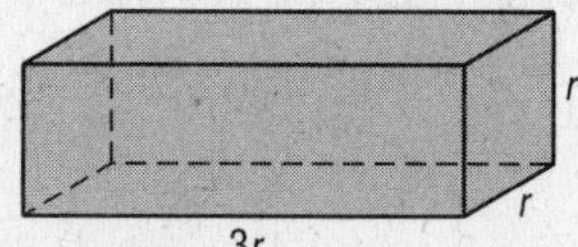

5.

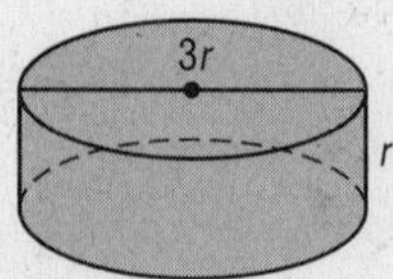

6.

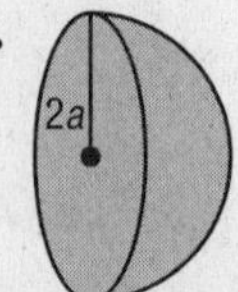

13-4 Study Guide and Intervention

Congruent and Similar Solids

Congruent or Similar Solids If the corresponding angles and sides of two solids are congruent, then the solids are congruent. Also, the corresponding faces are congruent and their surface areas and volumes are equal. Solids that have the same shape but are different sizes are **similar**. You can determine whether two solids are similar by comparing the ratio, or **scale factor**, of corresponding linear measurements.

Example **Describe each pair of solids.**

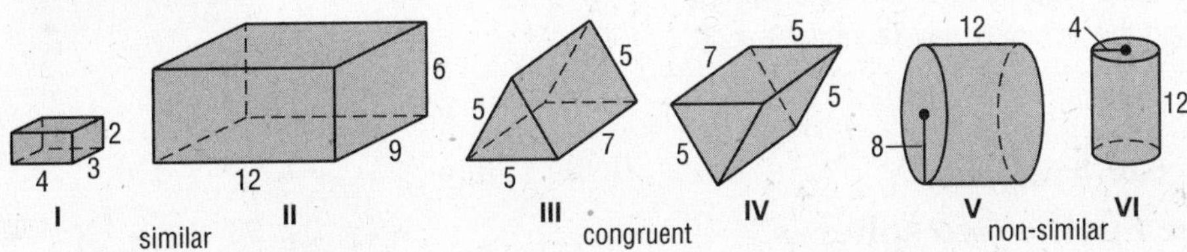

- Figures I and II are similar because the figures have the same shape. The ratio of each pair of corresponding sides is 1:3.
- Figures III and IV are congruent because they have the same shape and all corresponding measurements are the same.
- Figures V and VI are not congruent, and they are not similar because $\frac{4}{8} \neq \frac{12}{12}$.

Exercises

Determine whether each pair of solids are *similar*, *congruent*, or *neither*.

1.

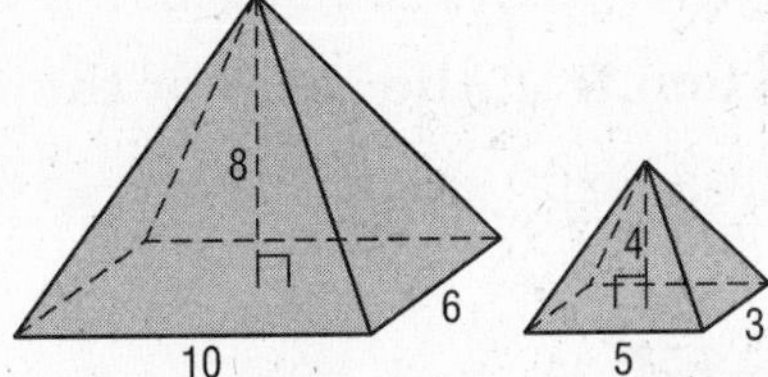

2.

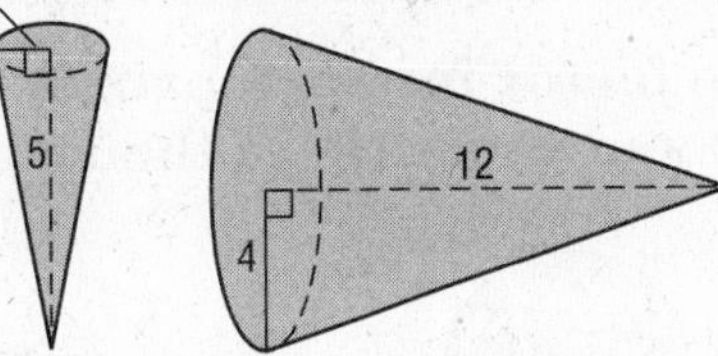

3.

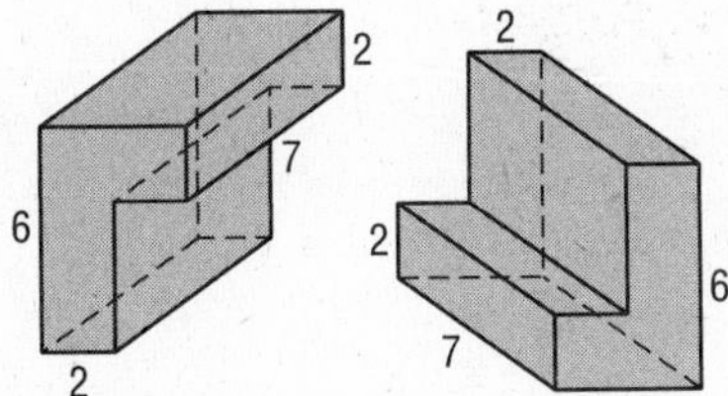

4.

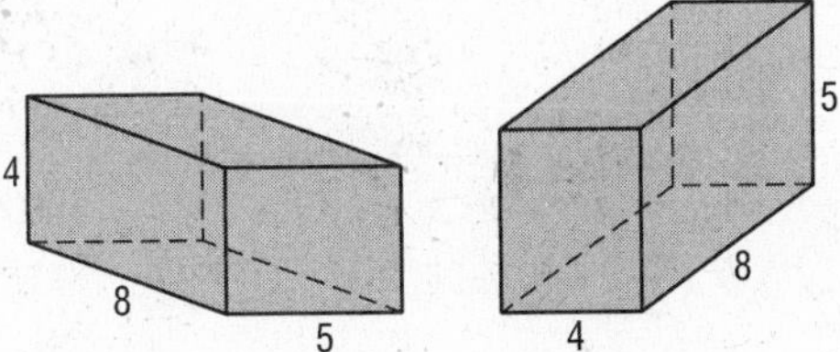

5.

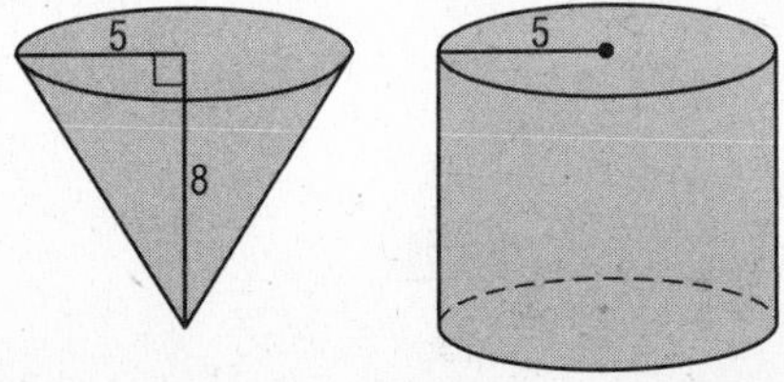

6.

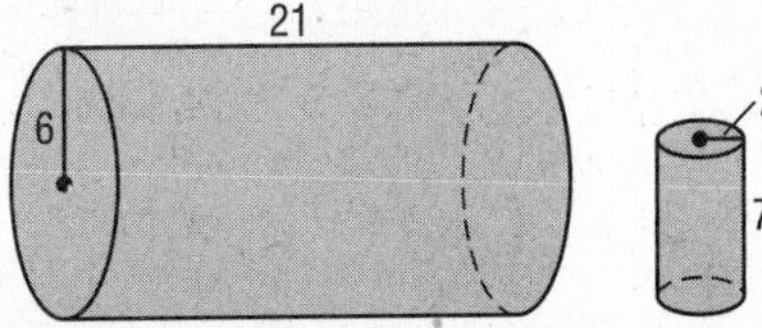

Lesson 13-4

13-4 Study Guide and Intervention *(continued)*

Congruent and Similar Solids

Properties of Similar Solids These two solids are similar with a scale factor of 1:2. The surface areas are 62 cm^2 and 248 cm^2 and the volumes are 30 cm^3 and 240 cm^3. Notice that the ratio of the surface areas is 62:248, which is 1:4 or $1^2:2^2$, and the ratio of the volumes is 30:240, which is 1:8 or $1^3:2^3$.

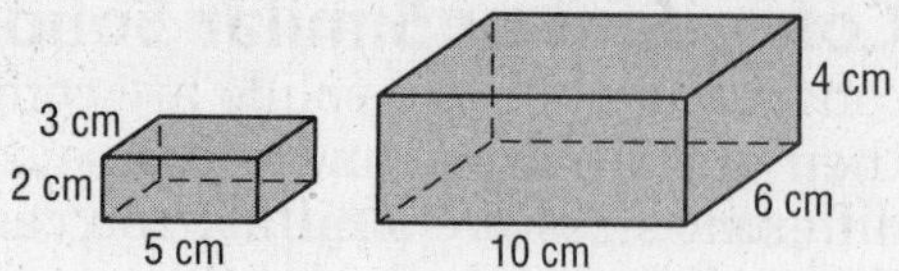

> If two solids are similar with a scale factor of $a:b$, then the surface areas have a ratio of $a^2:b^2$, and the volumes have a ratio of $a^3:b^3$.

Example **Use the two spheres.**

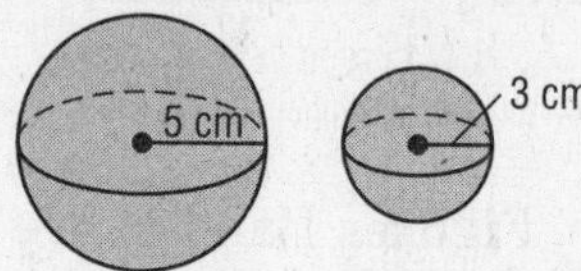

a. Find the scale factor for the two spheres.

The scale factor for the two spheres is the same as the ratio of their radii, or 5:3.

b. Find the ratio of the surface areas of the two spheres.

The ratio of the surface areas is $5^2:3^2$ or 25:9.

c. Find the ratio of the volumes of the two spheres.

The ratio of the volumes is $5^3:3^3$ or 125:27.

Exercises

Find the scale factor for each pair of similar figures. Then find the ratio of their surface areas and the ratio of their volumes.

1.

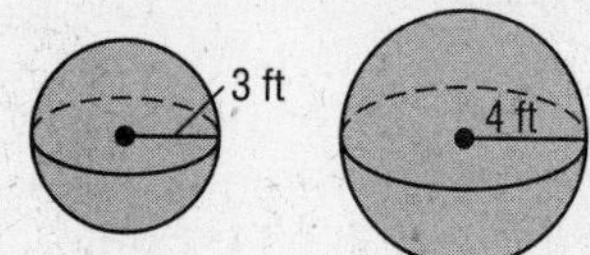

2.

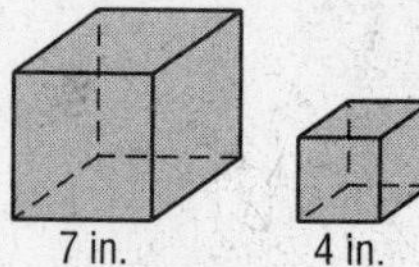

3.

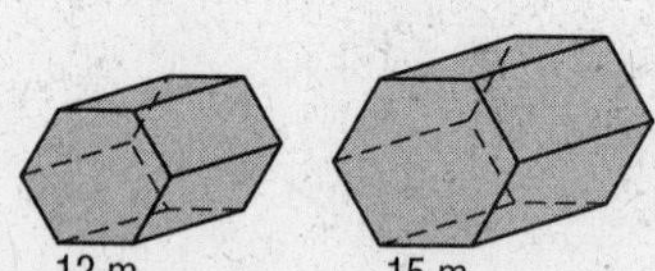

4.

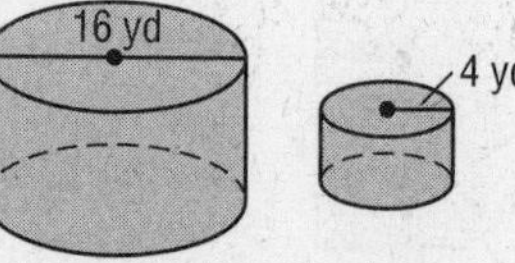

5.

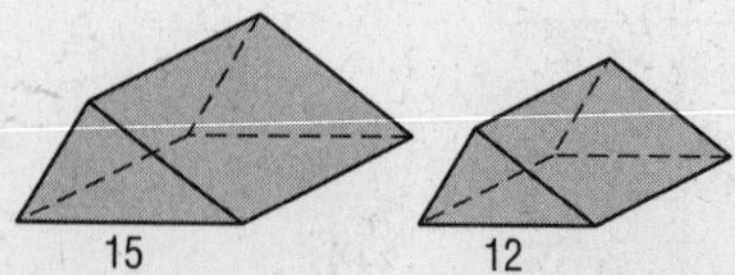

6.

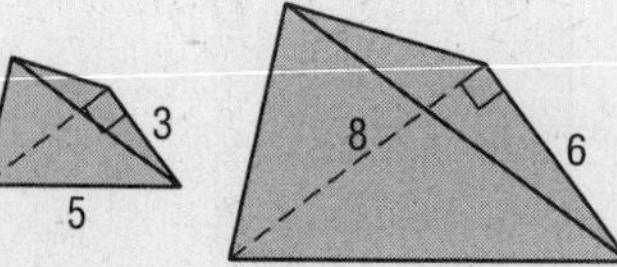

NAME ______________________ DATE __________ PERIOD _____

13-5 Study Guide and Intervention

Coordinates in Space

Graph Solids in Space In space, you can describe the location of a point using an **ordered triple** of real numbers. The x-, y-, and z-axes are perpendicular to each other, and the coordinates for point P are the ordered triple $(-4, 6, 5)$. A rectangular prism can be drawn to show perspective.

P(−4, 6, 5)

Example **Graph the rectangular solid that contains the ordered triple (2, 1, −2) and the origin. Label the coordinates of each vertex.**

- Plot the x-coordinate first. Draw a solid segment from the origin 2 units in the positive direction.
- Plot the y-coordinate next. Draw a solid segment 1 unit in the positive direction.
- Plot the z-coordinate next. Draw a solid segment 2 units in the negative direction.
- Draw the rectangular prism, using dotted lines for hidden edges of the prism.
- Label the coordinates of each vertex.

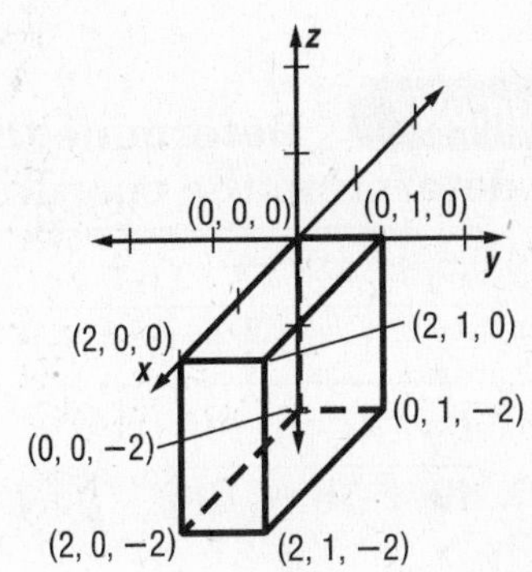

Exercises

Graph the rectangular solid that contains the given point and the origin as vertices. Label the coordinates of each vertex.

1. $A(2, 1, 3)$

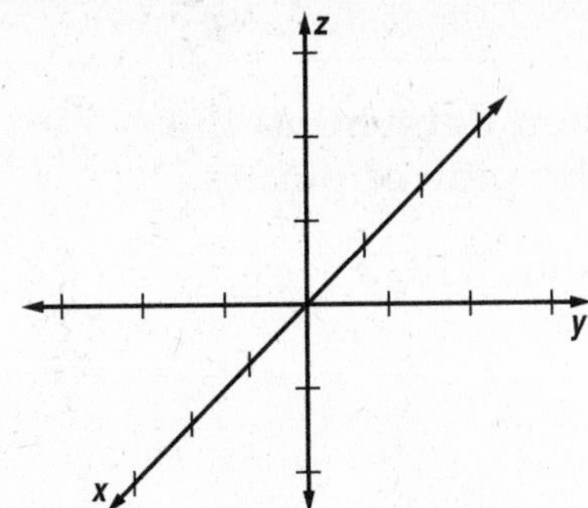

2. $G(-1, 2, 3)$

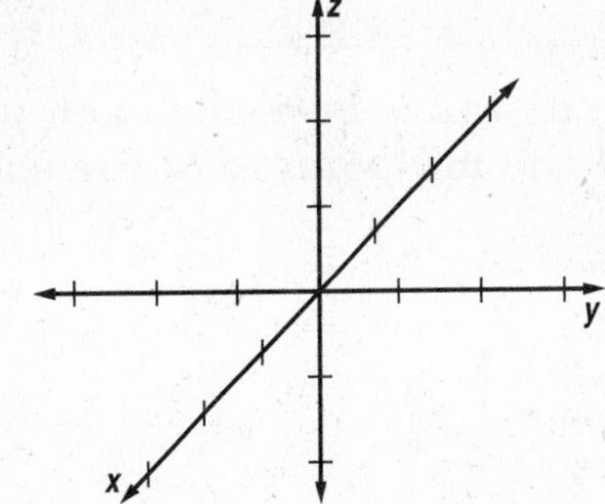

3. $P(-2, 1, -1)$

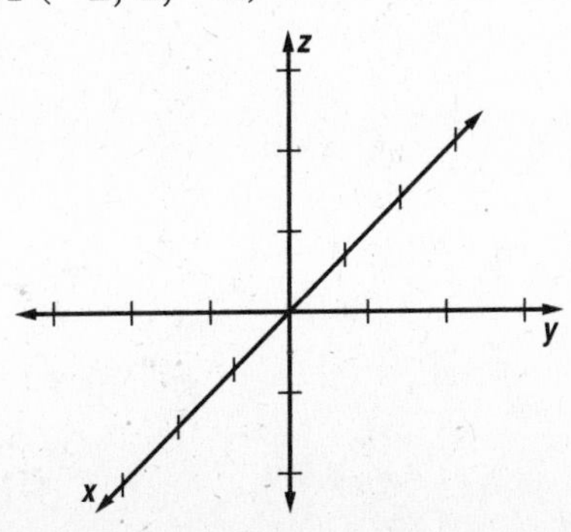

4. $T(-1, 3, 2)$

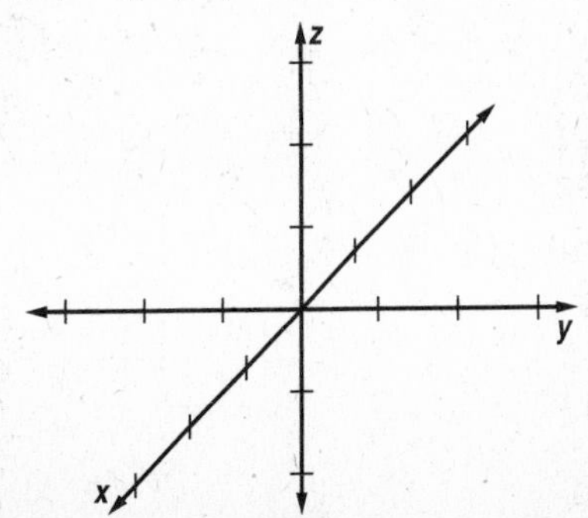

Lesson 13-5

13-5 Study Guide and Intervention *(continued)*

Coordinates in Space

Distance and Midpoint Formulas You can extend the Distance Formula and the Midpoint Formula to three dimensions to find the distance between two points in space and to find the midpoint of the segment connecting two points.

Distance Formula in Space	Given two points $A(x_1, y_1, z_1)$ and $B(x_2, y_2, z_2)$ in space, the distance between A and B is given by $AB = \sqrt{(x_1 - x_2)^2 + (y_1 - y_2)^2 + (z_1 - z_2)^2}$.
Midpoint Formula in Space	Given two points $A(x_1, y_1, z_1)$ and $B(x_2, y_2, z_2)$ in space, the midpoint of $\overline{AB}$ is at $\left(\frac{x_1 + x_2}{2}, \frac{y_1 + y_2}{2}, \frac{z_1 + z_2}{2}\right)$.

Example **Determine the distance between $A(3, 2, -5)$ and $B(-4, 6, 9)$. Then determine the coordinates of the midpoint of $\overline{AB}$.**

$$AB = \sqrt{(x_1 - x_2)^2 + (y_1 - y_2)^2 + (z_1 - z_2)^2}$$
$$= \sqrt{(3 - (-4))^2 + (2 - 6)^2 + (-5 - 9)^2}$$
$$= \sqrt{7^2 + (-4)^2 + (-14)^2}$$
$$= \sqrt{49 + 16 + 196}$$
$$\approx 16.2$$

$$\text{midpoint of } \overline{AB} = \left(\frac{x_1 + x_2}{2}, \frac{y_1 + y_2}{2}, \frac{z_1 + z_2}{2}\right)$$
$$= \left(\frac{3 + (-4)}{2}, \frac{2 + 6}{2}, \frac{-5 + 9}{2}\right)$$
$$= (-0.5, 4, 2)$$

Exercises

Determine the distance between each pair of points. Then determine the coordinates of the midpoint M of the segment joining the pair of points.

1. $A(0, 7, -4)$ and $B(-2, 8, 3)$

2. $C(-7, 6, 5)$ and $D(10, 2, -5)$

3. $E(3, 1, -2)$ and $F(-2, 3, 4)$

4. $G(-4, 1, 1)$ and $H(0, 2, -1)$

5. $J(6, 1, -2)$ and $K(-1, -2, 1)$

6. $L(-5, 0, -3)$ and $N(0, 0, -4)$